BASIC
ELECTRONICS
MATH

BASIC ELECTRONICS MATH

Clyde N. Herrick

Newnes

Boston   Oxford   Johannesburg   Melbourne   New Delhi   Singapore

Newnes is an imprint of Butterworth–Heinemann.

Copyright © 1997 by Butterworth–Heinemann

    A member of the Reed Elsevier group

Library of Congress Cataloging-in-Publication Data
Herrick, Clyde N.
    Basic electronics math / Clyde N. Herrick.
      p.   cm.
    Includes bibliographical references and index.
    ISBN 0-7506-9727-X
    1. Electronics—Mathematics.   I. Title.
TK7835.H39   1996
621.381′0151—dc20   96-38849
                                CIP

British Library Cataloguing-in-Publication Data
A catalogue record for this book is available from the British Library.

The publisher offers special discounts on bulk orders of this book.
For information, please contact:
Manager of Special Sales
Butterworth–Heinemann
313 Washington Street
Newton, MA 02158-1626
Tel: 617-928-2500
Fax: 617-928-2620

For information on all Newnes electronics publications available, contact our World Wide Web home page at: http://www.bh.com/bh/

10 9 8 7 6 5 4 3 2 1

Printed in the United States of America

Table of Contents

Preface

Electronics is a relative new and expanding field of scientific endeavor with many diverging areas, each changing rapidly, so that the devices we study today may soon be replaced by more sophisticated devices. There is however, a common core of knowledge that spans these many areas and that can be relied upon as the basis for understanding new electrical devices, circuits, and systems.

It is the purpose of *Basic Electronics Math* to provide the students of electronics with a sound background in mathematics and in mathematical concepts, as applied to the theory and analysis of electronic devices and circuits. This purpose is to be accomplished by treatment of the subject in a thorough and clear, yet simplified, manner. The student is lead by graduated exercises to a thorough comprehension of the mathematics of electronics.

Teaching methods have been improved in recent years, and it therefore becomes necessary that textbooks keep pace with advances in the technical sciences. This work has been prepared for the purpose of meeting this demand, and it is hoped that the material is appropriate to the methods of the most progressive electronics teachers.

The author wishes to acknowledge those who preceded him by preparation of other textbooks on this subject, and the many persons who aided him in the preparation of this work.

This text is respectfully dedicated as a teaching tool to the teachers of technical schools and community colleges.

Clyde N. Herrick
San Jose City College

Chapter 1

Arithmetic Fractions

 1.1   Introduction

The decimal number system is based on the fact that we have ten fingers. The digits are 0, 1, 2, 3, 4, 5, 6, 7, 8, and 9. Zero is the least significant number and 9 is the most significant number. When we reach ten in counting the one is carried over to the next column.

The place values are identified in the table below:

Table 1–1   Placement Identification of Decimal Numbers.

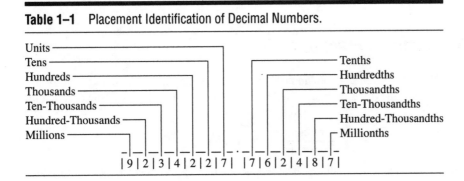

Decimal numbers are written after a period. For example: .01, .035, and .00045 are decimal fractions and are read one-hundredth, 35 one-thousandth, and 45 hundred-thousandths, respectively. A table of whole numbers and decimal fractions is shown above. For example: The number 9,234,227.762487 is read nine million, 2 hundred-thousand, three ten-thousand, 4 thousand, 2 hundred, twenty seven, seven tenth, 6 hundredth, 2 thousandths, 4 ten-thousandths, 8 one-hundred thousandths, and 7 millionths.

 1.2   Rounding Off Numbers

In technical mathematical problems we are often asked to round off numbers to some significant number of places. For example, a calculator might indicate an answer of 0.2378627 amperes for the current in an electrical circuit. However, typical laboratory instruments cannot measure to this accuracy. Furthermore, such accuracy is seldom necessary. We might be asked to *round off* the number to the third place, called the least significant digit (LSD). The rules specify that:

1. If the number after the specific round off point (LSD) is 5 or greater then the specific number should be increased by one.
2. If the number after the LSD is less than 5, the LSD remains the same.
3. Finally, if the LSD is even, it is increased by one if the next number is greater than 5.

Example 1 Round off to three significant numbers.

$$1237 = 1240$$
$$1234 = 1230$$
$$1253 = 1250$$
$$1255 = 1260$$
$$1245 = 1240$$
$$0.2456 = 0.2460$$
$$0.9873 = 0.9870$$

Exercises 1–1 Round off the following numbers to three significant digits.

1. 2346	**3.** 7235	**5.** 98750	**7.** 98764	**9.** 34650
2. 987.3	**4.** 0.1237	**6.** 0.01238	**8.** 2355	**10.** 52382

▶ 1.3 Common Fractions

A common fraction consists of a whole number above the dividing line, called a *numerator* and a whole number below the dividing line, called the *denominator*. The dividing line is called the *vinculum*.

$$\frac{\text{Numerator}}{\text{Denominator}} = \text{Fraction}$$

For example, $2/7$, $3/2$, and $1/2$ are fractions. The second fraction, $3/2$ is called an *improper fraction* because its value is greater than one. Fractions may have any value such as $1/1000$ or $2/2000000$. However we are prohibited from dividing by 0, as $1/0$ has no meaning.

▶ 1.4 Addition of Fractions

To add fractions with the same denominator, we add the numerators. For example: $1/7 + 3/7$ equals $4/7$. When the addition of fractions results in the numerator becoming greater than the denominator the result is an *improper fraction* that may be reduced to a mixed number.

$$\tfrac{1}{4} + \tfrac{3}{4} + \tfrac{1}{4} = \tfrac{5}{4} = \tfrac{4}{4} + \tfrac{1}{4} = 1\tfrac{1}{4}$$

To add or subtract fractions with different denominators we must change the fractions so that the denominators are of the same value. This action is called finding the *common denominator*.

$$\tfrac{1}{2} + \tfrac{1}{4} = \tfrac{2}{4} + \tfrac{1}{4} = \tfrac{3}{4}$$
$$\tfrac{1}{2} + \tfrac{1}{3} = \tfrac{3}{6} + \tfrac{2}{6} = \tfrac{5}{6}$$

A common denominator is the product of the denominators.

$$\tfrac{1}{2} + \tfrac{1}{3} + \tfrac{5}{6} =$$
$$\tfrac{18}{36} + \tfrac{12}{36} + \tfrac{30}{36} =$$
$$\tfrac{60}{36} = \tfrac{36}{36} + \tfrac{24}{36} = 1\tfrac{24}{36} = 1\tfrac{2}{3}$$

We note in the example above that the fractions $^{30}/_{36}$ can be reduced to a lower term. The necessity of this step can be prevented by finding the *lowest common denominator* (LCD) of the fractions. The LCD is the smallest number into which the denominators can be divided. To find the LCD in the example below, each denominator is changed to its prime numbers.

$$^1/_8 + ^1/_6 + ^5/_{12}$$

$$\frac{1}{2 \times 2 \times 2} + \frac{1}{2 \times 3} + \frac{5}{2 \times 2 \times 3}$$

The LCD is found by taking the product of the prime numbers in the denominator the greatest number of times that each appears in any denominator.

In the example above: 2 appears 3 times in $^1/_8$ and 3 appears only one time in $^1/_6$ and $^1/_{12}$. Therefore, the LCD is 24.

$$(2 \times 2 \times 2) \times 3 = 24$$

Then the fractions become:

$$\frac{1 \times 3}{3 \times 8} + \frac{1 \times 4}{4 \times 6} + \frac{2 \times 5}{2 \times 12} =$$

$$\frac{1 \times 3}{24} + \frac{1 \times 4}{24} + \frac{2 \times 5}{24} = \frac{3 + 4 + 10}{24} = \ ^{17}/_{24}$$

As another example:

$$\frac{2}{8} + \frac{5}{9} + \frac{7}{24} = \frac{2}{2 \times 2 \times 2} + \frac{5}{3 \times 3} + \frac{7}{2 \times 2 \times 2 \times 3}$$

$$\frac{2 + 5 + 7}{2 \times 2 \times 2 \times 3 \times 3} = \frac{18 + 40 + 21}{72} = \frac{79}{72} = 1\ ^7/_{72}$$

 ## 1.5   *Subtraction of Fractions*

Subtraction of fractions is accomplished in the same manner as addition. After finding the LCD of the fractions the smaller numerator is subtracted from the larger.

$$^1/_8 - ^1/_{12}$$

$$\frac{1}{2 \times 2 \times 2} - \frac{1}{2 \times 2 \times 3}$$

The LCD is $2 \times 2 \times 2 \times 3 = 24$

$$\frac{3}{24} - \frac{2}{24} = \ ^1/_{24}$$

 ## 1.6   *Mixed Numbers*

Mixed numbers are comprised of a whole number and a fraction. Examples of mixed numbers are $4\,^1/_2$ or $100\,^1/_4$. To add or subtract mixed numbers we may perform the function on the whole number and treat the fraction separately or change the mixed number to an improper fraction and perform the operation.

$$3\,^1/_2 + 2\,^3/_4 = 5 + ^1/_2 + ^3/_4 =$$

$$5 + ^2/_4 + ^3/_4 = 5\,^5/_4 = 6\,^1/_4$$

or

$$7/_2 + {}^{11}/_4 = {}^{14}/_4 + {}^{11}/_4 = {}^{25}/_4 = 6\,^1/_4$$

The operation to subtract mixed numbers is similar to addition. The numbers are changed to improper fractions and the smaller number is subtracted from the larger number.

$$5\,^1/_8 - 1\,^1/_2 = {}^{41}/_8 - {}^{12}/_8 = {}^{29}/_8 = 3\,^5/_8$$

Exercises 1–2 Solve the following problems.

1. $^1/_2 + {}^2/_3$
2. $^3/_4 + {}^7/_8$
3. $^2/_7 + {}^{15}/_{21}$
4. $3\,^1/_2 + 5\,^3/_4$

5. $5\,^7/_{21} + 3\,^5/_7$
6. $9\,^7/_{15} + {}^5/_8$
7. $8\,^2/_5 - 7\,^4/_{15}$
8. $9\,^1/_2 - 4\,^3/_4$

9. $15\,^1/_2 - 13\,^1/_4$
10. $^9/_5 - 2\,^1/_2$
11. $^3/_4 - {}^1/_2$
12. $4\,^{21}/_{22} - 7\,^3/_{11}$

 ## 1.7 Mathematical Expressions and Terms

The solution of mathematical problems involving addition, subtraction, division, and multiplication require specific sequences of operations. A *mathematical expression* is made up of mathematical *terms*. A term is a number proceeded by a + or − sign. An expression is comprised of one or more terms. For example, in the expression $5 - 3$, the $+5$ and the -3 are terms. The expression $5 \times 7 + 4$ and $8 \div 2 - 3$ are also comprised of two terms. The multiplication and division signs are within the term and must be acted upon first in each expression.

 ## 1.8 Signs of Grouping

To simplify the evaluation of mathematical expressions we use *signs of grouping*. The signs of grouping are:

- the parentheses $(2 + 3) \times (5 + 7)$
- the brackets $[4 + 5] \div [7 - 3]$
- the braces $\{7 - 3\} \times \{4 + 2\}$
- the vinculum $\overline{4 \times 5 + 3} - 2$ or $\overline{4 \times 5} + \overline{3 - 2}$

The rules for signs of grouping are that *all functions must first be performed within the grouping*.

$$(4 + 7) \times (3 - 1) = (11) \times (2) = 22$$

or

$$[(4 \times 3) - (8 \div 2)] = [(12) - (4)] = 8$$

The radical sign over a root is a sign of grouping

$$\sqrt{(120 + 8) \div (2 \times 4)} = \sqrt{(128) \div (8)} = \sqrt{16} = \pm 4$$

The correct sequence of operations within a grouping is multiplication, division, addition, and subtraction. The age old memory aid for the sequence operation is given in an 1810 math primer,

My Dear Aunt Sally, may help to solidify the process in your mind. An example is shown below. The proper use of signs of grouping will prevent any confusion.

$$7 - 6 \times 4 \div 6 = 7 - 24 \div 6 = 7 - 4 = 3$$
$$7 - [(6 \times 4) \div (6)]$$

Exercises 1–3 Solve the following problems:

1. $7 - 3 \times 6 + 2$

2. $14 \div 7 + 2 \times 5$

3. $22 + 5 \times 4 - 2$

4. $\frac{1}{2} \times \frac{2}{3} + \frac{6}{7} \div \frac{8}{21}$

5. $25 + 8 \times 3$

6. $125 \times 6 - 4 + 17$

7. $400 \div 10 - 17 + 7$

8. $20 \times 10 - 200 \div 20$

9. $(12 + 24) \times (13 - 5)$

10. $[2 \times (4 - 3) \div (82 - 56)]$

11. $\sqrt{9 \times (87 - 32) \div (33 - 31)}$

12. $[2 + 7\sqrt{22 - 60 + 42} - 5]$

Summary

1. Most electrical circuit applications allows for the rounding-off numbers to three significant places.
2. A common fraction always has a value of less than one.
3. An improper fraction always has a value of greater than one.
4. To add or subtract fractions a common denominator must be found for each of the fractions within the problem.
5. Mixed numbers are comprised of a whole number and a fraction.
6. The addition and subtraction of mixed numbers is best accomplished by changing each mixed number to an improper fraction and finding their common denominator.
7. Numbers are grouped to indicate the sequence of operations.
8. All operations within a grouping should be performed before the operation preceding the grouping is performed.

Chapter 2

Operations with Powers and Roots of Numbers

 2.1 Introduction

When we multiply a number by itself, we have performed an operation called *raising to a power*, and the product is called the *square of the number*. This terminology follows historically from the fact that the area of a square is equal to the length of one side multiplied by itself.

Example 1

$$2 \times 2 = 4$$

or

$$2 \text{ inches } \times 2 \text{ inches } = 4 \text{ square inches}$$

Thus, the numerical calculation is made with abstract numbers, and the concrete portion of the calculation is made with physical units. Before considering some examples, we should note that for brevity we use a symbol for the square of a number. This symbol is a raised 2 and is called an *exponent*. The number that is *raised to the power* is called the *base*. Thus, the equation $4 \times 4 = 16$ is usually written

$$4^2 = 16$$

Thus 4 is the *base*, 2 is the *exponent*, and 16 is the *square*.

An example of this concept to an electrical problem is the power dissipated in an electrical circuit. Power is the rate of doing physical work, or the rate of consuming electrical energy. Power values are measured in *watts*. We formulate the relations among voltage, current, resistance, and power as follows:

$$\text{Watts} = \text{volts} \times \text{amperes} = P = EI \tag{2.1}$$

$$\text{Watts} = \frac{\text{volts} \times \text{volts}}{\text{ohms}} = \frac{E^2}{R} \tag{2.2}$$

$$\text{Watts} = \text{amperes} \times \text{amperes} \times \text{ohms} = I^2 R \tag{2.3}$$

Formulas 2.1, 2.2, and 2.3 are equivalent to one another. We choose Formula 2.1 if a problem specifies voltage and current values and asks us to calculate the power value. Again, we choose Formula 2.2 if a problem specifies voltage and resistance values and asks us to calculate the power value. Or, if a problem specifies current and resistance values, and asks us to calculate the power value, we choose Formula 2.3, the quotient of two electrical units: amps and ohms.

If a number is multiplied by itself and the product multiplied again by the number, we say that the final product is the cube of the number, or the number raised to the third power.

Example 2

$$2 \times 2 \times 2 = 8 \text{ cubic units}$$

or

$$2^3 = 8 \text{ cubic units}$$

This operation is usually thought of as applying to a geometrical figure for volume. Of course, a cube may also apply to current temperature or other physical units.

If the cube of a number is multiplied by the number, we say the number has been raised to the fourth power as $x = y^4$. In this case, it is meaningless to let the number represent distance because space is three-dimensional. However, there are applications such as the radiation of heat which is four-dimensional. An example is Stefan-Boltzmann's law which states

$$\epsilon = ST^4 \tag{2.4}$$

where ϵ is the rate of heat radiation from a body, S is a constant, and T is the absolute temperature.*

Squares and cubes of numbers may be obtained from a calculator. To square a number enter it into the calculator and press x^2 button. To raise to a greater power the y^x button is used. For example, to find 2^5; enter 2 into the calculator, press 2, y^x, and enter 5 to yield the answer 32. Refer to your calculator instruction booklet for directions and the correct sequence of operations.

Exercises 2–1   Using a calculator, perform the following operations:

1. 3^3	**3.** 13^3	**5.** 5^3	**7.** 21^2	**9.** 7^4
2. 4^2	**4.** 11^2	**6.** 2^5	**8.** 13^3	

 ## 2.2   *Extraction of Square Roots*

When 3 is multiplied by itself, the product is 9, and we call 9 the square of 3. Note that 3 and 3 are the equal factors of 9. Thus, 3 is called the square root of 9. This particular square root is easy to calculate, or extract, because it is obvious from our knowledge of the multiplication table. For brevity, we use a symbol ($\sqrt{}$) to denote the square root of a number; this symbol is called a *radical sign*. Accordingly, we can write the example $3 \times 3 = 9$ in the form

$$3 = \sqrt{9}$$

which is read "3 equals the square root of 9."

Note that extraction of a square root is an *inverse* operation with respect to the squaring of a number. We find no difficulty in squaring a number, because we merely multiply the number by itself. On the other hand, extraction of a square root may appear impossible, unless we learn the rules for it. However, the scientific calculator makes the process simple if we may view the root of a number from another angle. Since the square root of 9 is 3 we may write the square root of 9 as either

$$\sqrt{9} \text{ or } 9^{1/2}$$

*Absolute temperature = 0 degrees Celsius temperature minus 273 degrees.

It is suggested by the author that the latter form always be used as will be demonstrated here and in a later chapter on Logarithms. For example: to find the square root of 121 to equal 11.

$$\sqrt{121} = (121)^{1/2}$$

Place 121 in the calculator press $\sqrt{}$ to read 11.

Again to find the fourth root of 444

$$(444)^{1/4} \approx 4.59$$

Place 444 in the calculator and press INV, $\sqrt[x]{}$, and 4 to read 4.59.

As an electrical application, let us consider Formula 2.5 in the example below. We are given the power and resistance values and are asked to calculate the current value. This operation entails the *extraction of a square root*. The *square root* of a number is one of its two *equal factors*. Observe the following example.

Example 3 The electrical power converted to heat in an electric coffee pot is 1200 watts. What is the current when the pot is connected into a 120 volt receptacle?

$$P = IE \tag{2.5}$$

Then

$$I = P/E = {}^{1200}\!/_{120} = 10 \approx 3.162 \text{ amperes}$$

 ## 2.3 Roots of Fractions

To extract the root of a fraction we can extract the root of the numerator and denominator separately, or we can make the division of the fraction into a decimal number and take the root of that number.

$$\frac{\sqrt{16}}{\sqrt{9}} = {}^{4}\!/_{3} \approx 1.333$$

or

$$\sqrt{\frac{16}{9}} = \sqrt{1.777} \approx 1.333$$

The simplest approach with the calculator is to divide ${}^{16}\!/_{9}$ and take the square root of the result.

 ## 2.4 Powers and Roots

A number or term is sometimes raised to both a power and a root.

$$(\sqrt{22})^3 = (22)^{3/2}$$

We may find the answer to this problem by taking the square root of 22 on our calculator and raising the answer to the third power. The process is as follows:

Place 22 in the calculator, press $\sqrt{x}$ for an answer of 4.6904, then press y^x, and 3. This yields the final answer of 103.189.

We will learn to use logarithms to find the answer to such problems in Chapter 16.

▶ 2.5 Higher Powers and Roots of Numbers

We have seen that the *root* of a number is one of its *equal factors*. Inversely, the power of a number is the number of times that the number is multiplied by itself. Of course, a number may have two equal factors or three equal factors.

Example 4

$$5 \times 5 = 25$$

Example 5

$$5 \times 5 \times 5 = 125$$

Just as 5 is the *square root* of 25, so is 5 the *cube root* of 125. Examples 10 and 11 may be written in the forms:

Example 6

$$\sqrt{25} = 5$$

Example 7

$$\sqrt[3]{125} = 5$$

The small 3 in Example 7 is called the *index* of the root. It is customary to omit the index 2 when we write the symbol for a square root. Of course, indexes higher than 2 must always be written. We never write the symbol for square root with an index of 1, because this would indicate merely a *redundant* calculation. In other words, $\sqrt[1]{5} = 5$.

In engineering work, we must sometimes solve problems entailing cube roots, fourth roots, and even higher roots.

Exercises 2–2 With your calculator, solve the following problems to three decimal places.

1. $(81)^{1/4}$
3. $\{(11)^{1/2}\}^5$
5. $(\sqrt{81})^5$
7. $[(8)^{1/3}]^5$
9. $9^{3/2}$

2. $(\sqrt{120})^3$
4. $(\sqrt{24})^4$
6. $(\sqrt{2})^3$
8. $[\{\sqrt{9}\}^3]^{1/4}$
10. $81^{5/2}$

Problems 2–1

1. What is the value of power dissipated in a light bulb drawing 1.2 A if its resistance is 81 ohms?
2. A 220 V oven dissipates 3,200 watts; what is its resistance?
3. What is the current through the heating element of the oven in Problem 2?
4. A 47 ½ watt soldering iron operates from a 6.3 volt source; what is its resistance?
5. A transistor radio operates from a 9 Volt source and draws 50 mA; what power value is being dissipated?
6. A color television receiver operating from a 115 V line has a power rating of 40 W; what is its effective resistance?
7. The starter on an automobile operates from a 12 V battery and consumes 600 W. Assuming no internal battery resistance and no lead resistance, what is the resistance of the starter?
8. What value of current is required by the starter in Problem 7?

Operations with Powers and Roots of Numbers   ▶ 9

9. A 12 V automobile battery supplies the following power values to six electrical devices:

head lights — 240 W
dash lights — 1.5 W
radio — 62 W
heater — 85 W
distributor — 52 W
tail lights — 25 W

What is the effective resistance value of each device?

10. What is the total value of current required to operate all the devices listed in Problem 9?

Summary

1. The roots of a number are two signed numbers ($\pm$). Usually only the positive number is valid in electronic circuit applications.
2. Operations with roots of numbers are best performed with a calculator and the $\sqrt{}$ function.
3. To extract the roots greater than 2, the $\sqrt[x]{y}$ function is utilized.
4. To raise a number to a power greater than 2, the calculator is best utilized with the y^x function.

Chapter 3

Scientific Notation and Powers of Ten

 3.1 Scientific Notation (Powers of Ten)

The electronics engineer and technician work with both very small and very large numbers. A pico farad capacitor (0.000,000,000,001 farad), a 1 megohms resistor (1,000,000 ohms), or the number of electrons to pass a point to produce an ampere of current (628 and 16 zeros added) are among the range of values encountered. These numbers are impractical to write out because of the length of time and large amount of space required. They are even more difficult to handle when used in mathematical operations. For these reasons, we will learn to write numbers in a standard form called *scientific notation.*

The employment of scientific notation for numbers requires an understanding of the *powers of ten. The power of a number is the number of times, less 1, that it is to be multiplied by itself.* As an example, 3 squared, or 3 to the second power, 3^2, means $3 \times 3 = 9$. Thus, 3^2 means that 3 is multiplied by itself once.

The *n*th power of the number 10 can be written

$$10^n = 10 \times 10 \times 10 \times \cdots \times 10 \tag{3.1}$$

The number 10 occurs (*n*) times. Thus, $10^3 = 10 \times 10 \times 10 = 1000$, 10 is the *base* and 3 is the *power* of the base called the *exponent.*

To place a number in the scientific notation form, the number is changed to a number between 1 and 10 with the proper power of 10. A few examples are given.

Example 1

$$2000 = 2 \times 10^3$$
$$28,900 = 2.89 \times 10^4$$
$$73 = 7.3 \times 10^1$$
$$0.025 = 2.5 \times 10^{-2}$$
$$1 = 1 \times 10^0 = 1$$

 3.2 Signs of Exponents

The sign of the exponent of the power of ten can be either positive or negative. A positive exponent or power indicates the number of zeros that are to be added to the integer, or the number of places to the right that the decimal point is to be moved. A negative exponent or power of ten indicates the number of places the decimal point is to be moved to the left.

Example 2

$$2.8 \times 10^3 = 2800$$

Example 3

$$2.8 \times 10^{-3} = 0.0028$$

The rules for converting numbers to scientific notation are:

1. *To change a large number to scientific notation, move the decimal point to the left to make the number read between 1 and 10. The number of places the decimal point is moved is the positive power of ten.*
2. *To change a decimal number to scientific notation, move the decimal point to the right to make the number read between 1 and 10. The number of places the decimal point is moved is the negative power of ten.*

Exercises 3–1 Change the numbers to their scientific notation with the proper power of ten:

1.	2700	**5.**	87,000	**9.**	0.000,000,01
2.	0.0113	**6.**	78,000	**10.**	6,280,000
3.	0.012	**7.**	0.000,012		
4.	126,000	**8.**	980,000,000		

 ## 3.3 Addition and Subtraction of Powers of Ten

To add or subtract numbers using powers of ten, all the numbers must be changed to have the same power of ten. This is very much like lining up all the decimal points when adding or subtracting decimal numbers.

Example 4

$$25 \times 10^3 + 24 \times 10^4 = 2.5 \times 10^4 + 24 \times 10^4 = 26.5 \times 10^4$$

This may be clearer if we remove the powers of ten. Then:

$$
\begin{array}{r}
25000 \\
240000 \\
\hline
265000
\end{array}
$$

Example 5

$$3.25 \times 10^4 + 1.2 \times 10^2 - 3.1 \times 10^{-2}$$

Again:

$$
\begin{array}{rl}
32500 + 120 - 0.031 = & 32500.000 \\
& 120.000 \\
& .031 \\
\hline
& 32619.969
\end{array}
$$

▶ 3.4 Multiplying with Powers of Ten

The rules for performing mathematical operations with powers of ten are the same as those for exponents. If you have not studied exponents in your previous mathematics courses, you will be able to apply the rules for the powers of ten when studying exponents in later chapters of this text.

Rule: *To multiply powers of ten, add the exponents.*

$$10^N \times 10^M = 10^{N+M}$$

Example 6

$$100,000 \times 100$$
$$100,000 = 1 \times 10^5 \text{ and } 100 = 10^2$$

then:

$$1 \times 10^5 \times 1 \times 10^2 = 1 \times 10^{5+2} = 10^7$$

Example 7

$$100,000 \times 0.0001$$
$$1 \times 10^5 \text{ and } 0.0001 = 1 \times 10^{-4}$$

then:

$$1 \times 10^5 \times 1 \times 10^{-4} = 1 \times 10^1 = 10$$

Example 8

$$20,000 \times 0.32$$
$$20,000 = 2 \times 10^4 \text{ and } 0.32 = 3.2 \times 10^{-1}$$

then:

$$2 \times 10^4 \times 3.2 \times 10^{-1} = 6.4 \times 10^3 = 6,400$$

Example 9

$$15,000 \times 0.00025 \times 30,000,000$$
$$15,000 = 1.5 \times 10^4, \quad 0.00025 = 2.5 \times 10^{-4} \quad \text{and} \quad 30,000,000 = 3 \times 10^7$$
$$1.5 \times 10^4 \times 2.5 \times 10^{-4} \times 3 \times 10^7 = 11.25 \times 10^7$$

▶ 3.5 Dividing by Powers of Ten

The law for dividing powers of ten can be summed by this rule. *To divide powers of ten, subtract the exponent of the denominator from the exponent of the numerator.* We may note here that powers of ten can be moved from the numerator to the denominator and from the denominator to the numerator by changing the sign.

$$\frac{1}{1,000} = \frac{1}{10^3} = 10^{-3} \tag{3.2}$$

$$\frac{1,000}{1} = \frac{1}{10^{-3}} \tag{3.3}$$

Example 10

$$\frac{100}{10} = 1 \times 10^2 \times 1 \times 10^{-1} = 10^{2-1} = 10$$

Example 11

$$\frac{250,000}{125} = \frac{2.5 \times 10^5}{1.25 \times 10^2} = 2 \times 10^{5-2}$$

Example 12

$$\frac{786,000}{0.0002} = \frac{7.86 \times 10^5}{2 \times 10^{-4}} = 3.93 \times 10^{5+4} = 3.93 \times 10^9$$

Powers of ten in the numerator and denominator with the same sign may be canceled. The author suggests that most powers of ten should be calculated mentally without the chance of an error in entering them into the calculator.

Example 13

$$\frac{2,000 \times 6,000}{100 \times 200 \times 300} = \frac{2 \times 10^3 \times 6 \times 10^3}{1 \times 10^2 \times 2 \times 10^2 \times 3 \times 10^2} = \frac{12 \times 10^6}{6 \times 10^6}$$

The 10^6 in the denominator cancels 10^6 from the numerator. Then dividing 12 by 6 yields:

$$^{12}\!/_6 = 2$$

 ### 3.6 Multiplication and Division Combined

Multiplication and division problems using powers of ten are solved by applying the individual rules of each, and by alternately multiplying and dividing.

Example 14

$$\frac{2,000 \times 0.0038}{0.000,005 \times 150} = \frac{2 \times 10^3 \times 3.8 \times 10^{-3}}{5 \times 10^{-6} \times 1.5 \times 10^2}$$

$$\frac{7.6}{7.5 \times 10^{-4}} = 1.013 \times 10^4$$

Exercises 3–2 Perform the following exercises with your calculator. Try eliminating the powers of ten mentally. State the answers in scientific notation.

1. $\dfrac{10^3 \times 10^{-4}}{10^4 \times 10^{-1}}$

2. $\dfrac{10^{-5} \times 10^7}{10^{-5} \times 10^6}$

3. $\dfrac{10^{-2} \times 10^8}{10^3 \times 10}$

4. $\dfrac{10^5 \times 10^{-3} \times 10^2}{10^3 \times 10^{-3} \times 10}$

5. $\dfrac{10^{-5} \times 10^2 \times 10^{-3}}{10^3 \times 10^{-2} \times 10^{-2}}$

6. $\dfrac{10^6 \times 10^{-4} \times 10^3 \times 10^{-4}}{10^{-7} \times 10^3 \times 10^4 \times 10^5}$

7. $\dfrac{234,000 \times 12}{0.000,22 \times 125}$

8. $\dfrac{0.009 \times 6.28}{746 \times 0.000,22}$

9. $\dfrac{982 \times 314}{0.000,56 \times 792}$

10. $\dfrac{67,000 \times 5,600}{6,450 \times 0.0037}$

11. $\dfrac{73 \times 456 \times 82,000}{0.000,32 \times 96,500}$

12. $\dfrac{8,760 \times 26 \times 2340}{456 \times 0.002 \times 0.3}$

3.7 Laws of Exponents

A summary of the laws of exponents is given below.

1. $10^2 \times 10^3 = 10^{2+3} = 10^5$

2. $10^5 \div 10^3 = 10^{5-3} = 10^2$

3. $(10^2)^3 = 10^{2\times3} = 10^6$

4. $(10^2 \times 10^3)^4 = 10^{2\times4} \times 10^{3\times4} = 10^{20}$

5. $10^8 \times 10^{12} = 10^{20}$

6. $\left(\dfrac{10^3}{10^2}\right)^4 = \dfrac{10^{12}}{10^8} = 10^{12-8} = 10^4$

3.8 Raising a Power of Ten to a Power

To raise a power of ten to a power, the exponent of the power of ten is multiplied by the power.

Example 15

$$(10^2)^3 = 10^{2\times3} = 10^6$$

Example 16

$$(10^{-3})^5 = 10^{-3\times5} = 10^{-15}$$

3.9 Taking the Root of a Power of Ten

To take the root of a power of ten, the exponent is divided by the power of the root.

Example 17

$$\sqrt[3]{(10)^6} = 10^{6/3} = 10^2$$

Perhaps a better approach is to change the power of the root to an exponent and use the raising to a power rule. Then Example 3-17 is solved as shown below.

Example 18

$$(10^6)^{1/3} = 10^{(6 \times 1/3)} = 10^{6/3} = 10^2$$

Problems 3-3

1. What is the value of the power dissipated in a heating element drawing 9.2 A if its resistance is 13 Ω?
2. A 220 V electric saw utilizes 2238 W, what is its effective resistance?
3. What is the current through the motor of the saw in Problem 2?
4. A 47 $\frac{1}{2}$ W soldering iron operates from a 120 V source; what is its effective resistance?
5. A portable telephone operates from 1 12 V source; what is its effective resistance?
6. A computer operates from a 120 V line and consumes 200 watts; what is its effective resistance?
7. The utility line for a house supplies the following power values to six electrical devices:

   Stove — 11 kW
   Dryer — 6.6 kW
   Washer — 746 W
   lights — 1.2 kW
   TV — 360 W
   Computer — 600 W

   What is the current and effective resistance of each device?

Express the following in scientific notation:

1. A 0.0001-F capacitor.
2. The wavelength of blue light, 0.000,047 cm.
3. The distance traveled by light in 1 second, 186,000 miles.
4. The number of seconds per year, 31,500,000.
5. The number of electrons in a coulomb of charge, 6,280,000,000,000,000,000.
6. The charge of an electron, 0.000,000,000,000,000,000,16.
7. Boltzmann's constant, 0.000,000,000,000,000,000,000,013,8.
8. The number of free carriers in germanium at room temperature, 24,000,000,000,000/cm^3.
9. The number of free carriers in silicon at room temperature, 20,000,000,000/cm^3.

Change the following to scientific notation with the proper power of ten and then to the proper prefix.

1. (a) 2,000,000,
   (b) 0.600,003,
   (c) 5,000,000,000,
   (d) .000,000,005,
   (e) 75,000,
   (f) 0.006,
   (g) 0.000,007,
   (h) 0.000,000,000,050

2. The AC reactance of an inductor, in ohms, is formulated $XC = 2\pi fL$ Ω; what is the value of the reactance XC when $\pi = 3.14$, $f = 10,000$ Hzs, and the inductance L is 20 μH?

Summary

1. The roots of a number are two signed numbers ($\pm$). Usually only the positive number is valid in electronic circuit applications.

2. Operations with roots of numbers are best performed with a calculator.
3. To extract the roots greater than 2, the $\sqrt[x]{y}$ function is utilized.
4. To raise a number to a power greater than 2, use the y^x function.
5. To simplify the mathematical operations pertaining to electronic circuits, it is important to utilize scientific notations with powers of ten.
6. To add or subtract numbers in scientific notation, each number must have the same value of power of ten.
7. To multiply numbers in scientific notation the numbers are multiplied and the powers of ten are added.
8. To divide numbers in scientific notation the numbers are divided and the power of ten of the denominator is subtracted from the power of ten of the numerator.
9. To raise numbers in scientific notation to a power, the numbers are raised to the power and the powers of ten are multiplied by the power.
10. To take the root of numbers in scientific notation the root is taken of the number and the power of ten is divided by the root.

Chapter 4

Units—Measurements and the Metric System

 4.1   Introduction

The solution to any practical mathematics problem entails a two-part answer. The first part represents the "how many," or the amount, and is always a number. This amount or magnitude is physically meaningless without the second part, which is the "what," or unit of the solution. In general, a unit is fixed by definition and is independent of physical conditions. Some examples of units are the foot, pound, degree, ohm, meter, and so on. Each of these is a *physical unit*, which means that it is a subject of *observation* and *measurement*.

Take, for example, the statement 115 volts; the number 115 is the amount or magnitude, and the term *volts* is the physical unit that gives the amount its meaning. Such numbers as "115 volts" are called *concrete numbers*.

The study of electronics will require you to learn many new terms. We will begin by defining some of the basic terms, and by reviewing the basic physical units.

 4.2   Units for Electronics

The Ampere   The ampere (amp or A) is the name given to the transfer of a certain number of electrons through a material over a certain elapsed time as a result of an electrical pressure. A movement of 6.25×10^{18} electrons (one coulomb) past a point in 1 second is defined to be 1 ampere of current.

The Electron Volt   The electron volt is used to state the energy of charged particles, such as electrons, and must not be confused with the volt unit. An electron which is accelerated through a potential difference of one volt gains one electron volt (ev) of energy.

The Volt   Voltage, electromotive force (emf), or potential difference is the electrical pressure, due to a charge separation, that forces an electron movement through a material. One volt is the electrical pressure necessary to cause 1 ampere of current to flow through 1 ohm of resistance. The symbol for a source voltage is (E), and the symbol for a potential drop due to a current flow is (V).

The Ohm   The Ohm (Ω) is the physical unit of resistance of a material. *One ohm* is the amount of electrical opposition, or resistance, that a material offers to limit current. *One volt across one ohm will result in a current of one ampere.* For example, the resistance of a copper wire 0.1 inch in diameter and 1,000 feet long is approximately 1 ohm.

The Siemen   The conductivity of a material is the ease with which it passes electrons. Conductivity and resistivity of a material are inversely related by the formula $(G = I/R)$. The unit of conductance is the Siemen. The symbol used for the Siemen is (S).

The Watt   The watt is the unit of power or the rate of doing work in an electrical circuit. The power converted in an electrical circuit is 1 watt when energy is converted at the rate of 1 joule-per-second.

The Coulomb   The coulomb (Q) is the unit of electrical charge. One coulomb is a charge of 6.25×10^{18} electrons. A flow of one coulomb past a point in one second is one ampere.

The Hertz   Frequency is the number of times an event occurs in a given period. In electrical circuits, frequency is usually given in *cycles-per-second*. By international agreement, the term Hertz (Hz) has been adopted to mean cycles-per-second. The National Bureau of Standards broadcasts standard frequencies at 400 and 440 kHz and 2.5, 5, 10, 15, 20, and 25 MHz on Radio stations WWV and WWVH.

The Second   The standard unit of time is the second. The Bureau of Standards maintains a frequency standard in which it is assumed that 9,192,631,770 cycles occur in one second.

 ## 4.3   *Ranges of Electrical Units*

As we noted earlier, electronics is a science which uses very large and very small units, such as a thousand-ohm resistor, a millionth-farad capacitor, a thousand million-cycles-per-second, and so on. To save time in writing and speaking these terms, symbols have been universally adopted to replace those most commonly used terms. Table 4-1 gives a list of terms and symbols.

Table 4–1   Table of the most common prefixes used in Electronics.

Prefix	Symbol	Value
Pico	p	10^{-12}
Nano	n	10^{-9}
Micro	μ	10^{-6}
Milli	m	10^{-3}
Kilo	k	10^{3}
Mega	M	10^{6}
Giga	G	10^{9}
Tera	T	10^{12}

Example 1

$$2,000 \text{ ohms} = 2 \times 10^3\ \Omega = 2 - \text{k}\Omega$$

or, as it is spoken, 2 kilohms or 2 "k" ohms.
   Again,

$$0.000,000,01 \text{ farad} = 0.01 \times 10^{-6}\ \text{F} = 0.01\ \mu\text{F} = 10 \text{ nF or } 10,000 \text{ pF}.$$

4.4   Systems of Measurement

Many systems of measurement are in use throughout the world. Most of these systems originally used parts of the human body as a reference. For instance, consider the foot in the English system. Historically, the foot was defined as the length of a certain king's shoe. These types of measurement are logical when we realize that this system was developed when there was no necessity for the great accuracy or uniformity of measurement that we have today.

The two systems of measurement with which we must become familiar are the metric system and the English system. The metric system, based on the decimal system, has as its references the meter for linear measurement and the gram for mass (weight). The English system has as its references the yard for the linear measure and the pound for mass (weight). The English system is used in the United States for most nonscientific measurements, and the metric system is used in most laboratory measurements. Most countries, including the United States, are converting to the metric system so that there is a standard in the world market.

 ## 4.5   The English System of Measurement

In the United States, the system of measurements of weights, lengths, volumes, frequency, and time are controlled by the Bureau of Standards in Washington D.C. We might note that the U.S. Congress has voted several times to have the metric system be the United States standard.

The standard unit of length in the English system is the yard, which is now based on the standard meter. The units of length in the English system and their relationship to each other follow in Table 4-2.

Table 4–2   English system of measurements and their relationships.

$$1000 \text{ milli-inches} = 1 \text{ inch}$$
$$12 \text{ inches} = 1 \text{ foot}$$
$$36 \text{ inches} = 1 \text{ yard}$$
$$3 \text{ feet} = 1 \text{ yard}$$
$$5\tfrac{1}{2} \text{ yards} = 1 \text{ rod, pole, or perch}$$
$$40 \text{ rods} = 1 \text{ furlong} = 220 \text{ yards}$$
$$5{,}280 \text{ feet} = 1 \text{ mile} = 1{,}760 \text{ yards}$$
$$8 \text{ furlongs} = 1 \text{ mile} = 1{,}760 \text{ yards}$$

For measurements at sea, the nautical mile of 6,080.21 feet is used; a speed of 1 nautical mile per hour is 1 knot.

 ## 4.6   The SI Metric System of Measurement

The metric system, based on multiples of ten, is the *International System of Weights and Measurements* based on the kilogram and meter. The International Measurements Society established the standard length of a meter as 1,650,763.73 wave lengths of orange-red radiation from isotope krypton 86.

The use of the metric system was made mandatory in all commercial transactions in France on July 4, 1837. It has gradually replaced other systems in all industrial nations of Western Europe and Asia, and has been made legal in the United States, Canada, Australia, and Great Britain.

The English system of measurement is very complicated and difficult to use and is becoming obsolete in the world market place. Teachers and students must make every effort to learn the metric system.

The units of the metric system are called the international system of units (SI) from the French "Le Systeme Internationale d'Unites." A summary of the most common SI units are given in Table 4-3.

Table 4–3   Basic Units of SI System.

Quantity	Unit	SI Symbol
Length	Meter	m
Mass	Kilogram	kg
Time	Second	s
Current	Ampere	A
Light intensity	Candela	cd
Molecular substance	Mole	mol
Thermodynamic temperature	Kelvin	K
Angle	Radian	rad
Capacitance	Farad	F
Conductance	Siemen	S
Electric charge	Coulomb	C
Frequency	Hertz	Hz
Flux density	Tesla	T
Force	Newton	N
Power	Watts	W
Inductance	Henry	H
Magnetic flux	Weber	Wb
Magnetic force	Ampere-turns	a°t
Resistance (Electric)	Ohm	Ω
Potential	Volts	V
Velocity	Meters per second	m/s
Temperature	Degrees Celsius	°C
Work (Energy)	Joule	J

 ## 4.7   A Summary of the Most Commonly Used Metric Measurements

A summary of the most commonly used metric measurements of length follow in Table 4-4.

Table 4–4   The Most Common Metric Measurements of Length

Length		
1,000 microns (μ)	=	1 millimeter (mm)
10 millimeters	=	1 centimeter (cm)
10 centimeters	=	1 decimeter (dec)
10 decimeters	=	1 meter (m)
10 meters	=	1 dekameter (dkm)
10 dekameters	=	1 hectometer (hm)
10 hectometers	=	1 kilometer (km)

A summary of the common metric measurements of weight is presented in Table 4-5.

Table 4–5   The Most Common Metric Weight Measurements.

Weight (Mass)		
10 milligrams (mg)	=	1 centigram (cg)
10 centigrams	=	1 decigram (dg)
10 decigrams	=	1 gram (g)
10 grams	=	1 dekagram (dkg)
10 dekagrams	=	1 hectogram (hg)
10 hectograms	=	1 kilogram (kg)

▶ 4.8 Relationships Between the Metric and the English Systems

The most often used units of linear measurement in the metric system are the millimeter, centimeter, meter, and kilometer. The relationship between these units and the units of the English system are as follows:

$$25.4 \text{ millimeters} = 1 \text{ inch (approximately)}$$
$$2.54 \text{ centimeters} = 1 \text{ inch (approximately)}$$
$$1 \text{ meter} = 39.37 \text{ inches (approximately)}$$
$$1 \text{ kilometer} = 0.62137 \text{ miles (approximately 5/8 mile)}$$

The milligram, the gram, and the kilogram are the most often used units of mass (weight) in the metric system. The relationship between these measurements and those of the English system are as follows:

$$1 \text{ milligram} = 0.0003527 \text{ ounces (approximately)}$$
$$1 \text{ gram} = 0.03527 \text{ ounces (approximately)}$$
$$1 \text{ kilogram} = 2.205 \text{ pounds (approximately)}$$

In electronics, measurements may be specified in either the English system or the metric system, and for this reason we must convert from one system to the other system of measurement. However, the metric system is preferred and should be used in all scientific writing. However, since both systems are in current usage we must learn to convert from one to the other.

Example 2   Light travels at a velocity of approximately 300,000 km per second. What is the approximate velocity of light in mi per sec?

$$1 \text{ km} \approx 0.62137 \text{ miles}$$

$$\frac{300,000 \text{ km/sec}}{0.62137 \text{ mi/km}} = 186,000 \text{ mi/sec (Answer)}$$

Example 3   How many kilograms does an 80 pound television set weigh? 1 kg = 2.205 lb.

$$\text{Weight in kg} = 2.205 \text{ kg/lb} \times 80 \text{ lb} = 176.4 \text{ kg (Answer)}$$

Example 4 How many inches are there in an antenna that has a length of 30 cm?

$$2.54 \text{ cm} = 1 \text{ in.}$$

$$\frac{30 \text{ cm}}{2.54 \text{ cm}} = 11.8 \text{ in. (Answer)}$$

Example 5 A football field measures 100 yards, what is the length in meters? (1 meter = 39.37 inches, one yard = 36 inches, therefore 1 meter = 1.0936 yards.)

$$100 \text{ yards} \times 1.0936 \text{ meters/yard} = 109.36 \text{ meters}$$

 4.9   The Micron and the Mil

The micron is a unit of length defined as one-thousandth part of a millimeter or one-millionth part of a meter (symbol μ). The mil is a unit of length defined as one-thousandth (1/1000) of an inch or 25.4001 microns. These units of length are used to describe the size of very small particles or very small movements of particles.

 4.10   Style and Usage of the International System of Units (SI)

The symbols for SI units and the convention which govern their use must be strictly followed to prevent confusion. Special care must be taken, to avoid misunderstanding, to use the correct case for symbols, units, and multiples (for example, K for Kelvin, k for kilo, m for meter or milli, M for Mega), and employ the correct punctuation. Wrong usage can cause serious misinterpretation. We will use the period (.) as the decimal marker (1.25) instead of the comma (1,25) which is used by some European countries. Table 4-3 should be consulted for your scientific writings.

The conversion tables in this chapter should be consulted for conversion factors between the systems. It is suggested that the tables in this chapter be copied and placed in your notebook for other subjects and future reference in your work as a technician.

Exercises 4–1 Complete the following problems.

1. 1,000,000 Ω =_____ $10^6 \ \Omega$ =_____ MΩ

2. 0.000,001 amp =_____ $\times 10^{-6}$ amp =_____ μamps

3. _____ Ω = $2.56 \times 10^4 \ \Omega$ =_____ kΩ

4. 0.000,000,028 F =_____ $\times 10^{-9}$ F =_____ nF

5. 3,250,000,000 Hz =_____ $\times 10^9$ Hz =_____ Ghz

6. 100 pF =_____ nF =_____ F

7. 220 μh =_____ mH =_____ H

8. 700×10^{18} coulombs =_____ m-coulombs =_____ T

Exercises 4-2 Convert the following measurements.

1. 39 cm = _____ in.
2. 82 km = _____ mi.
3. 36 in. = _____ cm.
4. 120 m = _____ cm.
5. 250 mils = _____ cm.
6. 8 yds = _____ mm.
7. 2 km = _____ yds.

8. 7 mi = _____ km.
9. 15 in. = _____ dm.
10. 5/8 in. = _____ cm.
11. 1/8 m = _____ cm.
12. 1/16 in. = _____ mm.
13. 120 μ = _____ mils.
14. 2500 mils = _____ in.

Problems 4-1

1. The length of a radio antenna is 30 in. What is its length in centimeters?
2. The wavelength of a television signal with a frequency of 77.25 MHz is 3.88 m. What is its wavelength in feet and inches?
3. The wavelength of a generator frequency of 15 kHz is 200 m. What is its length in yards, feet, and inches?
4. A radio wave travels at a velocity of 186,000 mi/sec and takes 0.015 sec to travel from New York to Hawaii. What is the distance in miles and kilometers?
5. The velocity of a sound wave at the normal room temperature of 68 degrees F is approximately 1,130 ft/sec. How far are you from a missile taking off if the blast is seen 6 sec before the sound is heard? Give your answer in feet and meters.
6. The frequency of green light is 576×10^{12} Hz and the corresponding wavelength is 20 μin. What is its wavelength in inches and centimeters? What is its frequency in kHz?
7. The frequency of red light is 430×10^{12} Hz and the corresponding wavelength is 0.000071 cm. What is its length in inches and micro inches? What is its frequency in tera Hertz?
8. The frequency of violet light is 732×10^{12} Hz (tera Hz) and the corresponding wavelength is 0.000016 in. What is its wavelength in centimeters?
9. The middle C note on a piano has a frequency of 256 Hz. What is the frequency of middle C in kHz?
10. The formula for the wavelenth of a given frequency is:

$$\text{Wavelength } (\tau) = 300 \times 10^6 / \text{frequency in Hz.}$$

    What is the wavelength of the frequency in Problem 9? in meters, in centimeters, in feet?
11. What is the wavelength of a 100 MHz sine wave in: meters, feet, centimeters?
12. What is the correct method of writing a frequency of 1,240,000,000 in metric notation?
13. The velocity of light is 300×10^6 meters per second; what is the velocity in feet per second?
14. A frequency on the citizen band is 22.7 MHz. Using the formula from Problem 10, determine the wavelength of that frequency.
15. A 360 cubic inch truck engine is _____ cubic centimeters.
16. What is the cubic inch rating of a 260 cubic centimeter automobile engine?
17. What is the exchange rate if an American dollar purchases 1.3 Australian dollars?
18. An often used sports saying is "the whole nine yards." What is this measurement in meters?
19. A football field is 100 yards. What is its length in meters?
20. An Australian football field is 30 meters in length. What is its length in yards?

Summary

1. The basic measurement of the English system is the inch. There are 2.54 cm per inch. This basic conversion can be utilized to make any distance conversion of metric to English or English to metric.
2. The basic measurement of weight in the metric system is the gram. One gram equals 0.03527 ounces. This basic conversion can be used to make any weight measurement between the metric and English system.
3. Electrical prefixes must be learned to understand the language of electronics.
4. The metric system is the foremost measurement system in industry.

Chapter 5

Computer Numbering Systems

 5.1 Introduction

The numbering system that has been developed for general usage is the base 10 system, probably because early man learned to count with his ten fingers. We utilize this system in our daily mathematics. The metric system and most monetary systems are based on this system.

The operation of computers is based upon the principles of semiconductor circuitry, or other materials that can be easily fabricated, and that have two states: off and on or high and low. These two states can also be thought of as a "1" or a "0." The numbering system developed for two state operations is called the *binary system* and is the language of the computer. The complexity of computers require higher order numbering systems such as the *octal (base 8)* and *hexadecimal (base 16)*. These are extensions of the binary numbering system.

 5.2 Decimal Numbering System (Base 10)

In the decimal numbering system each position can contain any of ten possible digits. These digits of course are: 0, 1, 2, 3, 4, 5, 6, 7, 8, or 9. Each digit in a multi-digit number has a weight factor based on a power of 10 (Chapter 3). In the five digit number 12345_{10} the (rightmost) digit will have a weighting factor of 10^0, the most significant digit (leftmost) has a weighting factor of 10^4 as shown in Example 1.

Example 1 The weight value of 5 digit decimal number.

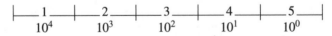

To determine the decimal number 12345, the digit of each position is multiplied by the appropriate weighting factor.

$$1 \times 10^4 + 2 \times 10^3 + 3 \times 10^2 + 4 \times 10^1 + 5 \times 10^0 =$$

$$
\begin{array}{r}
5 \\
40 \\
300 \\
2000 \\
10000 \\
\hline
\end{array}
$$

Answer   12,345

5.3   Binary Numbering System (Base 2)

As we mentioned earlier, digital electronics uses the binary numbering system because it uses only the digits 0 and 1 which can represent the *off* or *on*, *voltage* or *no voltage* condition of a circuit, or the *yes* or *no* conditions of other devices. For example, a magnetized element might represent a 1, while a nonmagnetized condition might represent a 0. Binary numbers have a weighting factor that can be represented similarly to a decimal number, as shown in Example 2.

Example 2   Position values of the first six positions of a binary number.

Place Value	MSD	5SD	4SD	3SD	2SD	LSD
	2^5	2^4	2^3	2^2	2^1	2^0
Decimal Equivalent	32	16	8	4	2	1

 ## 5.4   Binary to Decimal Conversion

Converting from binary (base 2) to decimal (base 10) is called *decoding* and is quite simple. We convert the digit in each position to decimal and add the results just as we did in the weighting of decimal numbers. This procedure is illustrated in Example 3.

Example 3   Convert 10101_2 to a decimal number.

Solution:

$$(1 \times 2^4) + (0 \times 2^3) + (1 \times 2^2) + (0 \times 2^1) + (1 \times 2^0) = 16 + 0 + 4 + 0 + 1 = 21_{10}$$

Since there are only 1's and 0's in the binary numbering system each digit position equals the weight of that digit position or it equals 0. The decimal values of the binary digit positions are added to give the decimal equivalent number. In the 2^4 position there is a 1. Therefore the decimal value is 16. In the 2^3 position there is a zero therefore the value is 0. In the 2^2 position there is a 1, therefore, the decimal value is 4. In the 2^1 position there is a 0, therefore, the value is 0. Finally, in the 2^0 position there is a 1, therefore, the value is 1. Conversions may be made with a calculator with a binary-decimal conversion feature.

Exercises 5–1   Convert the following binary numbers to their decimal values.

1. 1101	**4.** 1001	**7.** 10000	**10.** 10001
2. 1111	**5.** 10110	**8.** 11001	**11.** 1111001
3. 1000	**6.** 11111	**9.** 10111	

 ## 5.5   Decimal to Binary Conversion

Whenever an operator enters a decimal number into a digital computer, the computer must convert the number into the binary form of 1's and 0's called *bits*. This action is called *encoding*. Encoding is accomplished by repeated division of the decimal number by 2. If there is a remainder of 1 it is brought down as a binary 1 in the LSD of a binary number. If there is no remainder a 0 is indicated.

Example 4 Convert 75_{10} to binary.

$$73 \div 2 = 36 \qquad \text{Remainder of 1}$$
$$36 \div 2 = 18 \qquad \text{Remainder of 0}$$
$$18 \div 2 = 9 \qquad \text{Remainder of 0}$$
$$9 \div 2 = 4 \qquad \text{Remainder of 1}$$
$$4 \div 2 = 2 \qquad \text{Remainder of 0}$$
$$2 \div 2 = 1 \qquad \text{Remainder of 0}$$
$$1 \div 2 = 0 \qquad \text{Remainder of 1}$$
$$\text{Binary Number} = 1\,0\,0\,1\,0\,0\,1$$

The binary number is written bottom to top and right to left.

Exercises 5–2 Convert the following decimal numbers to binary numbers.

1.	1231	**3.**	122	**5.**	431	**7.**	21	**9.**	1872
2.	76	**4.**	673	**6.**	320	**8.**	16	**10.**	987

 ## 5.6 Octal Numbering System (Base 8)

The octal numbering system (base 8) has eight allowable digits 0, 1, 2, 3, 4, 5, 6, and 7. In counting the sequence is 0 to 7. On the next count we return to 0 as we do when we reach 9 in the decimal numbering system where a carry is generated. Table 5-2 on pg. 30 depicts a summary of numbering systems for binary, octal, hexadecimal, and decimal.

 ## 5.7 Converting Binary Numbers into Octal Numbers

Converting a binary number to an octal number is simply a matter of grouping the binary digits into groups of three starting from the LSD and moving to the left. Then writing down the octal (same as decimal) number for each group as shown in Table 5-1.

Table 5–1 Position value of the first four digits of an octal number.

MSD	3SD	2SD	LSD
8^3	8^2	8^1	8^0
512's	64's	8's	Units

Example 5 Convert 101001111 into an octal number.

$$101 \quad 001 \quad 111$$
$$5 \quad\;\; 1 \quad\;\; 7 \;= 517_8$$

The conversion of octal numbers to binary is simply the reverse of the preceding process. For example, 374_8 is converted to 011 111 100 in binary. The process is depicted in Example 6.

Example 6 Convert 2756_8 to binary.

$$
\begin{array}{cccc}
2 & 7 & 5 & 6 \\
010 & 111 & 101 & 110
\end{array}
$$

Exercises 5–3 Convert the following binary numbers to octal numbers. A scientific calculator with the octal option can be used for these operations.

1. 101111	**3.** 111111111	**5.** 101110011	**7.** 100111000
2. 110101	**4.** 101110001	**6.** 111000101011	

Convert the following octal numbers to binary.

1. 67_8	**3.** 777_8	**5.** 4652_8
2. 123_8	**4.** 137_8	**6.** 7654_8

 ## 5.8 Hexadecimal Numbering System (Base 16)

The hexadecimal number system (base 16) is used in computer systems that use a 16 bit or 32 bit code. The increase in the number of bits in the codes for computers was necessary as computers became increasingly more complex and required more memory.

The hexadecimal system follows the decimal system 0 through 9 and at that point converts to a letter system to represent each number as shown in Table 5-2. Note that a hexadecimal number may be identified by a 16 subscript or an H ($1BC_{16}$ or 1BCH).

One of the advantages of both the octal system and the hexadecimal system is that both make the reading and interpretation of binary numbers simpler for the human to interpret. Many digits are required to express a large binary number.

Example 7 Convert the binary number 110111001110 to an octal number, and to a hexadecimal number.

$$
\begin{array}{cccc}
(110 & 111 & 001 & 110)_2 \\
6 & 7 & 1 & 6_8
\end{array}
=
\begin{array}{ccc}
(1 & 1100 & 1110)_2 \\
1 & C & E_{16}
\end{array}
$$

Example 8 Convert $7EAD_8$ to a binary number.

$$
\begin{array}{cccc}
7 & E & A & D \\
0111 & 1110 & 1010 & 1101
\end{array}
= 111111010101101_2
$$

 ## 5.9 Addition of Binary Numbers

The addition of binary numbers is accomplished in the same manner as the addition of decimal numbers if we remember that the binary system has only 1's and 0's. The process is depicted in Example 9 with the addition of 1101 and 1011.

Table 5–2 Conversion Table for Binary, Decimal, Octal, Hexadecimal, and Binary-Coded Decimal.

Decimal	Binary	Octal	Hexadecimal	BCD
0	0000	0	0	0000
1	0001	1	1	0001
2	0010	2	2	0010
3	0011	3	3	0011
4	0100	4	4	0100
5	0101	5	5	0101
6	0110	6	6	0110
7	0111	7	7	0111
8	1000	10	8	1000
9	1001	11	9	1001
10	1010	12	A	0001 0000
11	1011	13	B	0001 0001
12	1100	14	C	0001 0010
13	1101	15	D	0001 0011
14	1110	16	E	0001 0100
15	1111	17	F	0001 0101
16	10000	20	10	0001 0110
17	10001	21	11	0001 0111
18	10010	22	12	0001 1000
19	10011	23	13	0001 1001
20	10100	24	14	0010 0000
21	10101	25	15	0010 0001
22	10110	26	16	0010 0010
23	10111	27	17	0010 0011
24	11000	30	18	0010 0100
25	11001	31	19	0010 0101
26	11010	32	1A	0010 0110
27	11011	33	1B	0010 0111
28	11100	34	1C	0010 1000
29	11101	35	1D	0010 1001
30	11110	36	1E	0011 0000
31	11111	37	1F	0011 0001
32	100000	40	20	0011 0010

Example 9

```
Carry     1          1    1
                 1    0    0    1
                 1    0    1    1
                ─────────────────
            1    0    1    0    0
```

The right most digits (2^0) are added to yield 10, or a 0 and a carry. The carry is moved over the next column (2^1) and added to yield again 10, or a 0 and a carry to the next column (2^2). This carry is added to the 2^2 column to yield a 1 with no carry. Finally, the 2^3 column is added to yield a 10 or a 0 and a carry to the 2^4 column. The carry is brought down to yield the total 10100. A carry is generated when the sum is greater than one in any column. The basic rules are:

1. $0 + 0 = 0$
2. $1 + 0 = 1$
3. $0 + 1 = 1$
4. $1 + 1 = 10$ or a 0 plus a carry of 1

Exercises 5–4 Add the following binary numbers. A scientific calculator with binary options can be used to perform these operations.

1. 1000
   1001

2. 1011
   1000

3. 1111
   1011

4. 0111
   1001

5. 1001
   1111

6. 1010
   1011

5.10 Addition of Octal Numbers

The addition of octal numbers follows the same rules as the addition of binary and decimal numbers, as illustrated in Example 10.

Example 10 Add the octal numbers 567 and 376.

Carry	1	1	1	
	5	6	7	
	3	7	6	
	(9)	(14)	(13)	
	−8	−8	−8	
1	1	6	5	$= 1175_8$

5.11 Additional of Hexadecimal Numbers

The addition of hexadecimal numbers is accomplished in the same manner as the addition of octal numbers as illustrated in Example 11.

Example 11 Add the hexadecimal numbers 4CF and A87.

Carry	1		1		
	4	C	F	
	A	8	7	
	(15)	(21)	(22)	
		16	16	
	F	5	6	F56H

5.12 Subtraction of Binary Numbers

Binary subtraction is accomplished in the same manner as decimal subtraction. We subtract from the LSD to the left. When a 1 is subtracted from a 0 we borrow from the next column.

Rules for Subtraction

1. $0 - 0 = 0$
2. $1 - 1 = 0$
3. $1 - 0 = 1$
4. $0 - 1 = 1$ (with a borrow from the next column)

Example 12   Step 1.

$$
\begin{array}{rcccc}
 & 1 & 0 & 1 & 0 \\
- & 0 & 1 & 1 & 1 \\
\hline
 & & & & 1
\end{array}
$$

subtract 1 from 0, borrow 1 from column 2

$$
\begin{array}{rcccc}
 & 1 & 0 & 0 & 0 \\
- & 0 & 1 & 1 & 1 \\
\hline
 & 0 & 0 & 1 & 1
\end{array}
$$

subtract 1 from 0 column 2, since column 2 and 3 are zero we must borrow from column 4.

This process is simple when using a calculator with binary functions. Select the binary mode, enter the larger number, press − and enter the smaller number.

Computers perform subtraction by a process of addition of complements. The one's complement for binary is found by changing all the 1's to 0's and all the 0's to 1's.

Example 13   Find the complement of a binary number.

$$1011001$$

Complement 0100110

To subtract one binary number the binary subtrahend is changed to a 1's complement and added to the minuend. If there is a carry left over from addition it is added to the sum of the numbers. This is called an *end around carry*. Example 14 illustrates this procedure.

Example 14   Subtract the binary numbers 1001110 from 1011001 by addition using the 1's complement.

(Minuend)	1011001		1011001	
(Subtrahend)	1001110		0110001	(one's complement)
		1	0001010	
			1	(end-around carry)
			1011	(difference)

Exercises 5–5   Perform the subtraction of the following binary numbers by the application of 1's complement.

1. 101
    100

3. 100
    010

5. 11101
    11011

2. 111
    101

4. 1101
    1011

6. 10111101
    10011110

▶ 5.13 Two's Complement

Another method of subtraction by computer is the *two's complement*. The two's complement of a binary number is found by adding a 1 to the one's complement. This procedure is shown in Example 15.

Example 15

$$11011011 \qquad \begin{array}{r} 00100100 \quad \text{(one's complement)} \\ +\qquad\quad 1 \\ \hline 00100101 \quad \text{(two's complement)} \end{array}$$

Subtraction is accomplished with the two's complement by adding the two's complement of the subtrahend to the minuend as shown in Example 16.

Example 16   Subtract the binary number 1000111 from the binary number 1100110.

$$\begin{array}{llll}
\text{(Minuend)} & 1100110 & 1100110 & 1100110 \\
\text{(Subtrahend)} & -1000111 & 0111000 & \\
& + & \underline{1 + 0111001} & \text{(two's complement)} \\
& & 1\ 0011111 & \text{(answer)}
\end{array}$$

The carry that results in the MSD in the answer is ignored. It indicates that the answer is a positive number. The number of digits or bits in the subtrahend must agree with the number of bits in the minuend. The lack of a carry in the MSD indicates a negative number as shown in Example 17.

Example 17   Subtract the binary number 110011 (51_{10}) from the binary number 100111 (39_{10}), to obtain an answer of -12_{10} or 1100. See Example 4.

$$\begin{array}{lllll}
100111 & (39_{10}) & 100111 & 100111 & (39_2) \\
-110011 & (51_{10}) & 001100 + 1 & \underline{001101} & \text{(two's complement)} \\
& & & 110100 & \text{(sum with no carry)}
\end{array}$$

Obviously, 110100 is not equal to -12. The lack of a carry indicates that the sum is negative and must be modified. This is accomplished by performing a two's complement to the sum.

$$\text{(Sum) } 110100 = 001011 + 1 = 001100 = (12_{10}) \text{ (answer)}$$

Exercises 5–6   Perform the indicated binary subtraction for the following.

1.	1101 -1000	**4.**	101101 -110000	**7.**	11001100 -10111101
2.	10111 -10011	**5.**	10110101 -10011101	**8.**	101110011 -110001010
3.	11110 -10111	**6.**	1000111 -1000010	**9.**	100100110 -011110001

5.14   Eight's and Sixteen's Complement

The eight's complement is found by taking the seven's complement and adding 1. The seven's complement is found by subtracting each digit of the octal number from seven as illustrated in Example 18.

Example 18   Find the eight's complement of 643_8.

$$777$$
$$\underline{643}$$
$$134 + 1 = 135 \text{ (eight's complement)}$$

The fifteen's complement of a hexadecimal number is found by subtracting each digit of the hexadecimal number from fifteen (F). The sixteen's complement is found by adding one to the fifteen's complement as shown in Example 5.19.

Example 19   Find the sixteen's complement of $C5A_{16}$.

$$FFF$$
$$\underline{C5A}$$
$$3A5 + 1 = 3A6 \text{ (sixteen's complement)}$$

5.15   Subtracting Octal Numbers

In the previous topics we discussed subtracting binary numbers by adding the two's complement of the subtrahend to the minuend.

Octal numbers are subtracted in computers by converting the subtrahend to the eight's complement and adding this to the minuend. This operation is shown in Example 20 where the octal number 567 (485_{10}) is subtracted from the octal number 745.

Example 20   Subtract the following octal numbers.

$$
\begin{array}{ll}
745 & \quad\quad\quad\quad 745 \\
\underline{-567} = 210 \text{ (7's complement)} + 1 & \quad \underline{+211} \\
\quad\quad\quad\quad \text{(Drop carry)} \quad 1 \quad 156 = (110110)_2 = 110_{10}
\end{array}
$$

Subtraction of hexadecimal numbers is accomplished by adding the sixteen's complement of the subtrahend to minuend. The sixteen's complement of a hexadecimal number is found by converting the number to the fifteen's complement and adding one. The process of subtracting hexadecimal numbers by the sixteen's complement is illustrated in Example 5.21.

Example 21   Subtract the hexadecimal number 177 (375_{10}) from 1E5 (485_{10}).

$$
\begin{array}{lll}
1E5 & FFF & \quad\quad\quad\quad 1E5 \\
\underline{-177} & \underline{-177} = E88 \text{ (15's Comp)} + 1 & \quad \underline{+E89} \quad \text{(sixteen's Comp)} \\
& \text{(Drop Carry)} \quad 1 \quad 06E \quad (110_{10})
\end{array}
$$

Exercises 5–7   Perform the indicated subtraction on the following octal numbers by the application of the eight's complement.

1.	123 −101	3.	641 −271	5.	6752 −5234	7.	3426 −1377
2.	723 −634	4.	111 −45	6.	7001 −6732	8.	3477 −2472

Perform the indicated subtraction of the following hexadecimal numbers.

1.	AC −9A	4.	EA −F2	7.	432 −3F7	10.	FFFF −2CF2
2.	1F −11	5.	1E2 −10C	8.	1234 −10CA		
3.	CF −A3	6.	3CA −269	9.	1CFD −1AA0		

Summary

1. Computers use the binary numbering system comprising of ones and zeros.
2. The octal system is a base eight system.
3. The hexadecimal system is a base sixteen system.
4. A byte is eight bits.

Chapter 6

Introduction to Algebra

▶ 6.1 Introduction

Although we might not have realized the fact, we employed principles of algebra in previous chapters. For example, by making use of our knowledge of operations with fractions, we learned that Ohm's law could be written in three forms. Thus, we observe that

$$E = IR \tag{6.1}$$

$$I = \frac{E}{R} \tag{6.2}$$

$$R = \frac{E}{I} \tag{6.3}$$

We recognize that instead of specifying voltage, current, and resistance values in whole numbers, mixed numbers, or fractions, we could use the symbols E, I, and R to denote *any* group of numerically related values. These are symbols for generalized numbers, and when we write the *algebraic formulas* we imply that IR stands for I multiplied by R.

We may progress to the formulation of expressions for power consumption. Observe that P, E, I, and R symbolize electrical units of power, voltage, current, and resistance in the following formulas.

$$P = EI \tag{6.4}$$

$$P = \frac{E^2}{R} \tag{6.5}$$

$$P = I^2 R \tag{6.6}$$

Moreover, Formula 6.4 states that *one* unit of power is equal to *one* unit of voltage, multiplied by *one* unit of current. Although the number 1 is not expressed, it is implied in each case. These formulas are a "short-hand" method of writing. Thus, we began to work with basic algebraic operations in considering groups of related numbers, numbers that are related by the fundamental operations of arithmetic.

▶ 6.2 Expressed and Implied Signs of Algebraic Operations

Recall that in the operations of arithmetic we employ signs of numerical operations.

Example 1

(a) $2 + 3 = 5$

(b) $3 - 2 = 1$

(c) $2 \cdot 4 = 2 \times 4 = 8$

(d) $3/2 = 3 \div 2 = 1.5$

(e) $5 \times 5 = 5^2 = 25$

(f) $\sqrt{81} = (81)^{1/2} = 9$

Many of the operational signs used in arithmetic are also used in algebraic equations and formulas. Others are not expressed, but implied. We shall also learn about new signs that imply various algebraic operations. Let us observe how Example 1 may be written algebraically.

Example 2

(a) $x + y = z$

(b) $x - y = z$

(c) $xy = x(y) = (x)(y) = z$

(d) $x/y = z$

(e) $\pm y^2 = x$

(f) $y^{1/2} = \sqrt{y} = \pm x$

Note in Example 2(c) that we omit the multiplication sign "$\times$," and merely write xy, which implies that x is to be multiplied by y. It would be confusing to write $x \times y = z$, because questions arise at whether "$\times$" symbolizes an abstract number or an algebraic operation. In the expression xy, x and y are called *factors* of the product. Again, we may denote the operation of multiplication by enclosing x, or y, or both, in *parentheses*. There is no advantage to be gained by employing parentheses in Example 2(c). On the other hand, we will find that it is essential to use parentheses when an equation entails the operations of both multiplication and addition (or subtraction).

Example 3

(a) $x(y + z) = w$

(b) $x(y - z) = w$

Example (a) and (b) imply either of the following operations:

(a)
$$xy + xz = w$$
x multiplied by $y + z = w$

(b)
$$xy - xz = w$$
x multiplied by $y - z = w$

Examples (a) and (b) do not imply that $xy + z = w$, nor that $xy - z = w$. In Example 2(f) the exponent $\frac{1}{2}$ symbolizes the radical sign for square root. This symbolization is entirely logical, as will be demonstrated subsequently.

Let us consider one more example of implied operations in algebraic formulation. Suppose a dry cell is connected in series with a switch and a lamp. When the switch is closed, current flows through the lamp filament, and electrical energy is commonly said to be consumed.

Strictly speaking, electrical energy is not consumed in the sense of being destroyed; instead, electrical energy is changed into an equal amount of heat energy plus light energy. Energy is equal to power multiplied by time, and is expressed in watt-seconds, or joules. Thus

$$\text{watt-seconds} = \text{volts} \times \text{amperes} \times \text{seconds}$$

$$W = EIt^* \text{ watt-seconds} \tag{6.7}$$

In Formula 6.7 the algebraic relation states that the number of watt-seconds, or joules, is equal to the product of the voltage, current, and time values. Thus, EIt symbolizes the product of these three values by simply writing the letters one after another.

▶ 6.3 Numerical and Literal Algebraic Expressions, Equations, and Formulas

A numerical algebraic expression comprises only numerals and signs of arithmetical operations. For example, $2(3 - 0.5) + 1.5$ is a numerical algebraic expression. On the other hand, a literal algebraic expression comprises generalized numbers, which may also be associated with numerals. Thus, E^2/R is a literal algebraic expression. Again, $P = E^2/R$ is an algebraic formula.

*Power $= E(Q/T)$ the rate of doing work. Work or energy is shown as $W = EIt$.

We know that if we write an expression such as $2 \times 3 \times 5$, each of these numbers is a factor of the product 30. Similarly, 2×3, 3×5, and 2×5 are factors of the product 30. If we write an expression such as $3xy$, it is called an algebraic expression, although the number 3 is present as a *coefficient*. In algebraic operations, any factor of a product is called a coefficient. Thus, 3, x, and y are each coefficients of the product $3xy$. Note that 3 is the *numerical* coefficient, and the x and y are the *literal* coefficients of the product. If a numerical coefficient is not expressed, 1 is always implied. For example, consider the product xyz; this product comprises the implied numerical coefficient 1, and the literal coefficients are x, y, and z.

▶ 6.4 Terms in Algebraic Expressions, Equations, and Formulas

Numbers that are separated by a plus or minus sign in an algebraic expression, equation, or formula are called *terms*. For example, let us consider the algebraic expression

$$E + IR - 0.5E = 0 \tag{6.8}$$

The terms in the expression are E, IR, and $0.5E$. Note that E is a *positive* term, because it is preceded by an implied plus sign. Again, IR is a positive term, because it is preceded by an expressed plus sign. However, $0.5E$ is a *negative* term, because it is preceded by an expressed minus sign.

Next, let us consider the algebraic formula

$$\frac{0.2E}{I} + \frac{0.3e}{I} + R = 1 + R \tag{6.9}$$

The terms in this formula are $0.2E/I$, $0.3e/I$, and R. If we subtract R from both sides of this formula, we obtain

$$0.2\frac{E}{I} + 0.3\frac{e}{I} + R - R = 1 + R - R$$

$$0.2\frac{E}{I} + 0.3\frac{e}{I} = 1$$

In other words, a formula or equation can be compared to a balance; if we add or subtract equal weights to or from each side of the balance, the balanced condition will be maintained. Since R is separated by a minus sign in the preceding formula, we recognize that R is a term in the formula.

▶ 6.5 Similar Terms

Similar terms, or *like terms*, are terms that are *exactly* alike in their literal parts. For example, EI, $-0.5EI$, and $89EI$ are similar terms. On the other hand, *dissimilar terms*, or *unlike terms*, are *not* exactly alike in their *literal* parts. Thus, EI, eI, EIt, and E^2i are dissimilar terms. Not that terms such as EI, ei, and Ei, and eI are dissimilar terms because upper-case and lower-case letters are used to denote voltages and currents in different parts of a circuit. Therefore, in spite of the superficial similarity of such terms, they are in fact dissimilar terms.

6.6   Monomials, Binomials, Trinomials, and Polynomials

A *monomial* is an algebraic expression that has only one term. For example, EIt is a monomial. A *binomial* is an algebraic expression that has two terms; thus, $IR + Ir$ is a binomial. A *trinomial* is an algebraic expression that has three terms; for example, $x/a - by + cz$ is a trinomial. A *polynomial* is an algebraic expression that has two or more terms; thus, binomials and trinomials are also polynomials.

6.7   Subscript and Prime Notations

We have seen that if we are concerned with two voltages and two currents in a circuit, the voltage might be denoted E and e, and the currents might be denoted I_1 and I_2. Again, if we wish to distinguish between two resistors in a circuit, one resistance might be identified as R and the other as r. *Subscripts* are used in a similar manner to distinguish between individual voltages, currents, resistances, powers, energies, and so on. Subscripts may be either numerical or literal. For example, three individual voltages might be denoted E_1, E_2, and E_3; or E_a, E_o, and E_e; or E_A, E_B, and E_C. This notation is read "E sub one," "E sub two," and "E sub three," or "E sub a," "E sub b," and "E sub c." In the foregoing example, it is preferred to read "E sub small a," "E sub small b," and "E sub small c," to clearly distinguish this group from "E sub large A," "E sub large B," and "E sub large C."

Note that R_1^2 is read "R sub one, squared," and denotes that value R_1 is to be multiplied by itself. Similarly, R_2^2 is read "R sub two, squared," and denotes that the value R_2 is to be multiplied by itself. To put it another way, subscripts are used to identify the resistance with which we are concerned, and exponents are used to denote algebraic operations.

Instead of subscript notation, *prime* notation is often used to identify various voltages, currents, resistances, and so on. For example, power values occurring at two different points in a circuit may be denoted P and P'. Again, if a third power value is to be considered, it may be denoted P''. The symbol P' is read "P prime," and the symbol P'' is read "P double-prime" or "P second." Although it is permissible to write P'^2, it is generally preferred to write P_1^2. Of course, these notations denote $P' \times P'$, and $P_1 \times P_1$. If a fourth power value is to be considered, we may write P''', which is read "P triple prime."

6.8   Substitution of Numerical Values and Evolution

An algebraic expression is equal to a numerical value that can be calculated, or *evaluated*, when the numerical values of the letters in the expression are given. To evaluate the algebraic expression, we must substitute the values of the letters in the expression, and then perform the indicated algebraic operations.

Example 4   Consider the circuit and data given in Figure 6-1. To calculate the value of I, we observe the I_2 is related to I_1 and I_3 by Kirchhoff's current law,

$$I_2 = I_1 - I_3 \tag{6.10}$$

The value of I_1 is given, but the value of I_3 must be calculated. We perceive that the value of I_3 is related to E_3 and R_3 by Ohm's law.

$$I_3 = \frac{E_3}{R_3} \tag{6.11}$$

Since $E_3 = E_2 = 3V$, and $R_3 = 1.5\,\Omega$, we calculate that $I_3 = 2\,A$. Next, we substitute the known values of I_3 and I_1 in Formula 6.10, and solve for I_2.

$$I_2 = 6 - 2 = 1 \text{ amp}$$

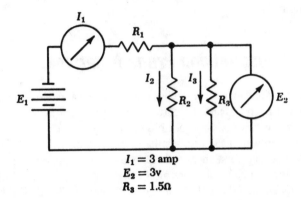

Figure 6–1
A circuit in which the value
of I_2 is to be calculated.

$I_1 = 3$ amp
$E_2 = 3v$
$R_3 = 1.5\Omega$

Problems 6–1

1. Explain the distinction between expressed and implied signs of algebraic operations.
2. Explain the distinction between a numerical and a literal algebraic expression.
3. Explain the distinction between similar terms and dissimilar terms.
4. Identify each of the following expressions as a monomial, binomial, trinomial, and/or polynomial.

6.9 Algebraic Addition

The introductory section of this chapter confronted us with the problems of addition and subtraction with signed numbers. We observed how these problems are approached. Now we will proceed to develop general rules for algebraic operations with signed numbers. The general rules for addition follow easily from consideration of problems entailing debts and assets.

Example 5 Suppose you have $100 in the bank, and deposit another $50. Then your bank account becomes

Addend,	$100
Addend,	$50
Sum,	$150 (Answer)

Example 6 On the other hand, suppose that your bank account is overdrawn $75. In this case, your bank account becomes −$75, which means that you owe the bank $75. Next, if you overdraw another $50, your bank account becomes

Addend,	−$ 75
Addend,	− 50
Sum,	−$125 (Answer)

Thus, it follows that the rule for addition of signed numbers that have the same signs is: *Calculate the sum of the absolute values of the addends, and write their common sign before the sum.*

Example 7   Next, suppose that your bank account is overdrawn $30, and that you deposit $75; then your bank account becomes

Addend,	−$ 30
Addend,	+ 75
Sum,	+$ 45 (Answer)

It follows that the rule for addition of signed numbers that have different signs is: *Calculate the difference of the absolute values of the addends, and write the sign of the number with the largest absolute value before the sum.*

Example 8   Again, suppose that you have a bank account of $175 on Monday, write a check for $200 on Tuesday, deposit $100 on Wednesday, deposit $50 on Thursday, and write a check for $250 on Friday. Your bank account on Friday becomes

Addend,	+$ 175
Addend,	− 200
Addend,	+ 100
Addend,	+ 50
Addend,	− 250
Sum,	−$ 125 (Answer)

When several signed (algebraic) numbers are to be added, as in Example 9, the operation is simplified by first calculating the subtotal of the positive addends, then calculating the subtotal of the negative addends, and then calculating the sum of the subtotals.

Example 9

+ $175	− $200
+ 100	− 250
+ 50	− $450
+ $325	

Subtotal,	− $450
Subtotal,	+ 325
Sum,	− $125 (Answer)

▶ **6.10   Algebraic Subtraction**

The general rule for subtraction of signed numbers is easily understood by considering some typical examples. To subtract, as we know, means to *deduct* from a number. If we subtract 3 from 4, we write $4 - 3 = 1$. We read "Subtract 3 from 4." Note that we subtracted 3 by *changing the sign* of $+3$ *and adding* -3 to 4. We write

Example 10

Number,	4	Minuend
Subtract,	−3	Subtrahend
	1	Difference

Restatement of the problem:

$$\begin{array}{ll} \text{Addend,} & 4 \\ \text{Addend,} & \underline{-3} \\ & 1 \end{array}$$

In other words, if we *change the sign of the subtrahend* in a subtraction problem, we obtain an equivalent problem in addition, and the answer is found by *algebraic addition* of the addends. We shall find that this principle holds true also if we subtract a negative number from another number. For example, with reference to a dB meter that indicates -2 dB. It is obvious that if we should add 2 dB to -2 dB, the reading would become zero. On the other hand, if we should subtract 2 dB from -2 dB, the reading will become -4 dB.

Similarly, the 0 degree point on a thermometer scale represents an arbitrary assignment. For example water freezes at 0 degrees on the Celsius scale; on the other hand, this same absolute temperature corresponds to 32 degrees on the Fahrenheit scale. On the *absolute*, or Kelvin scale, 0 degrees corresponds to absence of heat, or zero heat energy. We should note that water freezes at 273.04 degrees on the absolute or Kelvin scale. Thus, 32 degrees F, 0 degrees C, and 273.04 degrees K corresponds to the same energy state. Obviously, *negative Kelvin temperatures have no physical existence.*

Signed concrete numbers also denote values with respect to an arbitrary zero point in an electric circuit. For example, consider the electrical *ground* in an electrical circuit. A ground may be a connection to earth, as a connection to a water pipe that runs down into the soil; again, a ground may be a connection to the metal framework of an electrical assembly. In either case, voltage values are customarily measured with *respect to ground*.

To add or subtract numbers they must be in the same units. We cannot, for example, add resistors and capacitors, unless we are to form a new category called parts.

Example 11   Subtract 72 centimeters from 3.6 meters. We must change meters to centimeters or centimeters to meters in order to perform the operation.

$$3.6 \text{ meters} = 360 \text{ centimeters} - 72 \text{ centimeters} = 288 \text{ cm or } 2.88 \text{ m}$$

 ## 6.11 Addition of Polynomials

To add polynomials, we arrange the addends in columns that have like terms, and then add the individual columns.

Example 12   Let us add $ax^2 + bx + c$ to $2bx + 3c + 4ax^2$. We proceed as follows:

$$\begin{array}{ll} \text{Addend,} & ax^2 + bx + c \\ \text{Addend,} & \underline{4ax^2 + 2bx + c} \\ \text{Sum,} & 5ax^2 + 3bx + 2c \quad \text{(Answer)} \end{array}$$

Suppose that $a = 1$, $x = 2$, $b = 3$, and $c = 4$. Then we substitue and evaluate the sum.

$$\begin{array}{ll} \text{Addend,} & 4 + 6 + 4 \\ \text{Addend,} & \underline{16 + 12 + 4} \\ \text{Sum,} & 20 + 18 + 8 \quad = 46 \text{ (Answer)} \end{array}$$

6.12 Subtraction of Polynomials

To *subtract* polynomials, we proceed in the same manner as for addition, except that we change *all* signs in the subtrahend, and then add algebraically.

Example 13 Let us subtract $ax^2 + bx + c$ from $4ax^2 + 2bx + 3c$. We write

Addend,	$4ax^2 + 2bx + 3c$
Subtrahend,	$-\ ax^2 -\ bx -\ c$
Sum,	$3ax^2 +\ bx + 2c$ (Answer)

Problems 6–2

1. A man had 150 horses and sold 46. He then used the money he received to purchase 72 calves. How many head of livestock does he own?
2. An automotive storage battery has the following inputs and outputs: output to lights, 45 amp; output to the radio, 6 amp; output to the ignition, 5.5 amp; and input from the generator, 51 amp. What value of charge or discharge does the ammeter indicate?
3. A boy tries to run up the down escalator; the escalator's absolute speed is 80 yd/sec, and the boy's absolute running speed is 200 ft/sec. What is the boy's actual speed and direction?
4. A jet plane travels at an absolute speed of 600 mph into a 150 mph headwind. What is the plane's ground speed?
5. The outside temperature is -5 degrees F; the inside temperature is 67 degrees F. What is the temperature differential?
6. The ambient temperature around a light bulb is 25 degrees C. What is the difference in temperature if the filament is operating at 3000 degrees F?
7. The line used in a tug-of-war game has the following applied forces; end A has 121 lb, 87 lb, 102 lb, and 67 lb applied; end B has 97 lb, 102 lb, and 67 lb applied. Which direction will the line move, and with what force?
8. Construct a number scale from 0 to $+10$ and 0 to -10. On this scale, count the following units: 3 right, 5 left, 9 right, 8 right, 7 left, 2 right, and 10 left. What is the final point on the number scale?
9. A jet airliner traveling from San Francisco to New York moves at an air speed of 835 ft per second. What is the airliner's speed in relation to the earth, which rotates at a speed of 1,035 mph?
10. What is the speed of the jet airliner in Problem 9 relative to earth on the return trip from New York?
11. A generator's output into a circuit consisting of two motors is 1,119 W. One of the motors is rated at 1 horsepower. What is the energy conversion in W sec for the second motor? (Note: 1 hp = 746 W.)
12. A decibel meter is set up to read a reference of 0 dB at the output of an amplifier. With two new frequencies the readings are $+5.2$ dB and -1.5 dB, respectively. What is the difference in dB gain between these two measurements?
13. A man's bank balance is $152.88 before he makes the following transactions: deposits, $192.75; write check, $17.50; cash withdrawal, $246.00; deposits, $27.99; deposits, $78.72; writes check, $175.86; and writes check, $129.33. What is his bank balance?
14. What is the total decimal gain* of an amplifier if the gains of each section are -0.5 dB, 30 dB, -3.2 dB, 29 dB, and -1.8 dB?

*Decibel gain dB — add for total gain.

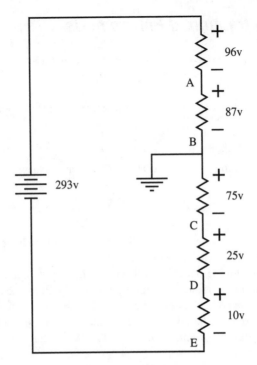

Figure 6–2
Series electrical circuit that
develops both negative
and positive voltages.

15. With respect to Figure 6-2, what is the voltage at

(a) point B with respect to point C?
(b) point A with respect to point D?
(c) point E with respect to point D?
(d) point C with respect to point E?
(e) point C with respect to point A?
(f) point A with respect to point E?
(g) point E with respect to point A?

Summary

1. The equal sign in an equation indicates that one side of the equation is equal to the other side.
2. An equation can be thought of as a balance in which one side can be modified in any way if the same modification is made to the other side.
3. When symbols of grouping enclose other symbols of grouping, the innermost symbols of grouping must be removed first.
4. A positive number is any number greater than zero.
5. A negative number is any number less than zero.
6. To add two quantities of like signs, add their absolute value and prefix the sum with their common sign.
7. To add two quantities with opposite signs, find their difference and prefix the sign of the quantity with the greatest value.
8. To subtract two signed quantities change the sign of the subtrahend and proceed as with addition.
9. To add polynomials, arrange the addends in columns of like terms, and add the individual columns.

10. To subtract polynomials, proceed as with addition, but change the signs of all terms in the subtrahend.
11. When symbols of a grouping are preceded with a plus sign, the symbols may be removed without a change of sign.
12. When symbols of groupings are preceded by a minus sign, the removal of the symbol of groupings must be accompanied with a change of signs of all enclosed terms.

Chapter 7

Algebraic Multiplication of Monomials and Polynomials

 7.1 Introduction

We have observed elementary principles of algebraic multiplication in preceding discussions. Now we are in a good position to consider this operation in greater detail. Multiplication is a process for calculating the sum of several numbers that are all alike. Thus, 2 multiplied by 3 yields a product that is equal to the sum of 2 plus 2 plus 2. Of course, this sum is equal to the product of 3 multiplied by 2, or 3 plus 3. Again, 1/2 multiplied by 4 yields a product that is equal to 1/2 plus 1/2 plus 1/2 plus 1/2. Although it is more difficult to visualize that 4 multiplied by 1/2 is the calculation of 4 added one-half time, it is helpful to note the following:

Example 1

   (a) $4 \times 4 = 4 + 4 + 4 + 4 = 16$
   (b) $4 \times 2 = 4 + 4 + 0 + 0 = 8$
   (c) $4 \times 3/2 = 2 + 2 + 2 + 0 = 6$
   (d) $4 \times 1/2 = 2 + 0 + 0 + 0 = 2$

The progression in Example 1 provides an introduction to the meaning of 4 added one-half time. In turn, this meaning becomes quite clear when we observe that $4 = 1 + 1 + 1 + 1$.

 7.2 Multiplication of Signed Numbers

We have become familiar with both positive and negative numbers, and have previously noted some basic principles of operations with these signed numbers. Now we will develop some general rules for the algebraic multiplication of signed numbers. This development is facilitated by asking what is meant by multiplication of signed numbers from the viewpoint of addition.

Example 2 If we multiply 2×3, we may write

   (a) $+2 + 2 + 2 = +6$

Again, if we multiply -2×3, we may write

   (b) $-2 - 2 - 2 = -6$

Example 2(a) is obvious. On the other hand, it is not obvious that $3(-2) = -6$. What are we doing when we add 3 minus-two times? To answer this question, it is helpful to consider numbers that represent distances along a straight line. We assign the number zero as a *reference point* on the line; this point separates the line into intervals that have positive and negative *directions*. Negatively-directed distances are measured with negative numbers, and positively-directed numbers are measured with positive numbers.

This discussion leads us to the general rules for algebraic multiplication of signed numbers.

1. *The product of two terms with like signs is positive.*
2. *The product of two terms with unlike signs is negative.*
3. *If more than two signed factors are multiplied together, the sign of the product is found by repeated multiplication of the -1 and $+1$ factors which are expressed or implied.*

To make certain we clearly understand Rule 3, let us observe the following operations.

Example 3 $-4 \times +5 \times -3 = (-4 \times +5) \times -3 = -20 \times -3 = +60$

$-4 \times -5 \times -3 = (-4 \times -5) \times -3 = +20 \times -3 = -60$

$\begin{aligned}
&+ \text{ expressed,} && (+x)(-y) = -xy \\
&+ \text{ implied,} && x(-y) = -xy \\
&+ \text{ implied,} && (-x)(y) = -xy \\
&+ \text{ expressed,} && (-x)(-y) = +xy \\
&+ \text{ implied,} && (-x)(-y) = xy \\
& && (x)(y)(z) = xyz
\end{aligned}$

The application of signed numbers is important in electrical circuits. For example, voltage is normally measured with respect to *ground* or a common point called ground. Figure 7-1 illustrates a circuit in which the voltage values are both negative and positive with respect to ground.

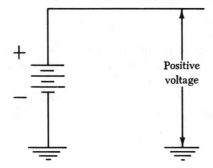

The voltage is positive with respect to ground.

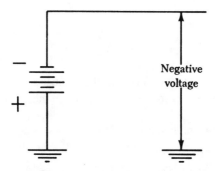

Figure 7–1
Voltages measured to ground.

The voltage is negative with respect to ground.

Exercises 7–1 Multiply the following:

1. $(-2)(3)$
2. $(-3)(-6)$
3. $x(-y)$
4. $5(-6)$
5. $3(-4)5$
6. $-x(y)(-z)$
7. $-5(x - y)$
8. $-x(y + z)$
9. $3(x - y)$
10. $-35(16 - x)$
11. $(a)(-b)(-c)(-y)$
12. $-12(-8)(a)(b)(-c)$

▶ 7.3 Multiplication of Signed Numbers with Exponents

We have noted some elementary principles of operations with exponents in previous discussions of powers of ten. For example, a number that is raised to a power is called the *base*, and the power to which it is raised is called the *exponent*.

The general rule for multiplication of numbers with exponents when the base is the same follows.

4. *To calculate the product of numbers with exponents when the base is the same, express the base to a power that is the sum of the exponents.*

Observe the following equations.

$$RR^2 = R^{1+2} = R^3$$
$$EE = E^{1+1} = E^2$$
$$R^3R^{-1} = R^{3-1} = R^2$$

An exception to Rule 4 is observed when the base is zero.

$$0^2 \times 0^3 = 0^5 = 0$$

This equation represents a trivial operation because it is of little mathematical worth or importance. Its only significance is as an illustration of the logic and consistency that governs mathematical operations.

Next, let us consider the powers of negative numbers. If we square the number -3, we must multiply -3 by -3, and its square is evidently 9.

Example 4

$$(-3)^2 = 9$$

Again, if we cube the number -3, we must multiply -3 by -3 and then multiply the product by -3.

Example 5

$$(-3)^3 = -27$$

The parentheses in Examples 4 and 5 indicate that the number enclosed within the parentheses is to be multiplied by itself. The sign before the parentheses in these examples is an implied plus sign.

Now, let us consider the following notation.

Example 6

$$-(3^2) = -9$$

Example 6 states that the number enclosed within the parentheses is to be multiplied by itself, and then multiplied by -1.

Example 7

$$-(-3^3) = -(-27) = 27$$

We proceed in each case to perform the operations in the parentheses first, and then complete the remaining calculations.

Exercises 7–2   Perform the indicated operations:

1. $(-3)^3$
2. $-x(x)^2$
3. $x(x)$
4. $x^2 \cdot x^3 \cdot x^5$
5. $I^2 \times I^2$
6. $a(E)^2(E)^3$
7. $-(-2)^3$
8. $(-a)^2 \times a^3$
9. $(-4ac^2)^2$
10. $-(5a^2c)^3$
11. $(2b^3)^2$
12. $y \cdot y^3$

 7.4   Multiplication of Monomials Containing Exponents

The foregoing section has laid the groundwork for multiplication of monomials that contain exponents. Let us illustrate this procedure in the examples below.

Example 8

$$2x^2y \times 3xy^3 = 6x^3y^4$$

Again, let us multiply $-2a^3x^2$ by $4ax^3$.

Example 9

$$-2a^3x^2 \times 4ax^3 = -8a^4x^5$$

These examples lead us to the general rule for multiplication of monomials that contain exponents.

5. *Calculate the product of the numerical coefficients (expressed or implied), and write the resulting sign before the product in accordance with the rules for algebraic multiplication.*
6. *Multiply the signed product of the numerical coefficients by the product of the literal factors, with observance of the law of exponents.*

 7.5   Multiplication of Polynomials by Monomials

We are now in a position to develop the general rule for this algebraic operation. Let us consider an example of such application to an electrical problem. With reference to Figure 7-1, we know from Kirchhoff's voltage law that $E = E_1 + E_2$. Ohm's law states that $E_1 = IR_1$, and the $E_2 = IR_2$. Ohm's law also states that $I = E/R_T$, where R_T is the total resistance in the circuit; thus,

$$R_T = R_1 + R_2$$

In turn, we may write

$$E = IR_T = I(R_1 + R_2) = IR_1 + IR_2$$

Note that I is a monomial, and $(R_1 + R_2)$ is a binomial. Accordingly, the product $IR_1 + IR_2$ is obtained from multiplication of a binomial by a monomial.

Again, let us consider the power relations in the circuit of Figure 7-2.

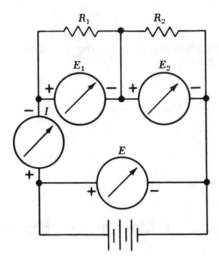

Figure 7–2
Analysis of this circuit illustrates the multiplication of a binomial by a monomial.

The power dissipated in R_1 is formulated $P_1 = I^2 R_1$, and the power dissipated in R_2 is formulated $P_2 = I^2 R_2$. The power supplied by the battery is equal to the sum of the power values dissipated in R_1 and R_2. In turn, we may write

$$P_T = P_1 + P_2$$
$$P_T = I^2 R_1 + I^2 R_2$$

This example leads us to the general rule for algebraic multiplication of polynomials by a monomial.

7. *To multiply a polynomial by a monomial algebraically, multiply each term in the polynomial by the monomial, observing the law of exponents, and write the proper sign before each term in the product.*

 ### 7.6 *Multiplication of a Polynomial by Another Polynomial*

The algebraic multiplication of a polynomial by another polynomial entails calculating the product of the multiplicand and each term in the multiplier, and calculating the sum of similar terms.

Example 10   Multiply $I_1 R_1 - I_1 R_2$ by $I_2 R_1 + I_2 R_2$.

$$
\begin{array}{ll}
\text{Multiplicand,} & I_1 R_1 - I_1 R_2 \\
\text{Multiplier,} & \underline{I_2 R_1 + I_2 R_2} \\
& I_1 I_2 R_1^2 - I_1 I_2 R_1 R_2 \\
& \underline{I_1 I_2 R_1 R_2 - I_1 I_2 R_2^2} \\
\text{Product,} & I_1 I_2 R_1^2 - I_1 I_2 R_2^2 \quad \text{(Answer)}
\end{array}
$$

It may happen that the sum of similar terms is zero.

Example 11

$$\begin{array}{ll}
\text{Multiplicand,} & 2ax + 2by \\
\text{Multiplier,} & \underline{2ax - 2by} \\
& 4a^2x^2 + 4abxy \\
& \quad\quad\quad \underline{- 4abxy - 4b^2y^2} \\
\text{Product,} & 4a^2x^2 - 4b^2y^2 \quad \text{(Answer)}
\end{array}$$

Again, it may happen that there are no similar terms.

Example 12

$$\begin{array}{ll}
\text{Multiplicand,} & 2a^2x + 2by \\
\text{Multiplier,} & \underline{3a - b} \\
\text{Product,} & 6a^3x + 6aby - 2a^2bx - 2b^2y \quad \text{(Answer)}
\end{array}$$

These demonstrations lead us to the general rule for multiplication of a polynomial by another polynomial.

8. *To multiply polynomials algebraically, calculate the product of the multiplicand and each term in the multiplier, and calculate the algebraic sum of any similar terms that may result.*

Note that it makes no difference which polynomial is regarded as the multiplicand and which polynomial is regarded as the multiplier. If we are given three polynomials, and are required to find their product, we choose any two of the polynomials and calculate their product; then we multiply this product by the third polynomial to obtain the answer. Any number of polynomials may be multiplied in this manner.

Exercises 7–3 Multiply the following polynomials and monomials:

1. $(8x)(2xy)$
2. $(-3ab)(a + b)$
3. $(2b^2)(a - b)$
4. $5abc(-4ab)$
5. $(-2kh)(-8khm)$
6. $(a^2m)(an)(-a^3mn)$
7. $(-3ax^2)(-7a^2x^3)$
8. $-5(4p' - 5p'')$
9. $8I_1(R_1 - R_2)$
10. $n_2(z_1 + z_2)$
11. $2xy^2(1 - 3x + 2x^2 - 3x^2)$
12. $(-3L)(-4Lm)(L + m)$
13. $\frac{1}{2}x^2(10ax - 6x^2y)$
14. $4x^2(2ax - 9a^2x^2)$

15. $(a + b)(a - b)$
16. $(m + n)(m - y)$
17. $(h + k)(h - k)$
18. $(x^2 + y^2)(x^2 - y^2)$
19. $(2x + 3y)(5x + 7y)$
20. $(a - 1)(a^2 + 2ab + b^2)$
21. $(x + 2)(x^4 - 2x^3 + 6x^2 + x - 1)$
22. $2a[6 - 3a(a^2 - 2)]$
23. $(2x^2 + 3x + 1)(x^2 + 2x + 1)$
24. $(12x^5 + y^6)(2x^7 - y^8)$
25. $(I_1 - I_2)(2R_1 + 3R_2 + 4R_3)$
26. $(\frac{1}{2}a^2 - \frac{1}{3}x^2)(2a^3 + 6x^3)$
27. $(a^{n+1} + a^3b)(a^{n-1} + a^2b^2)$
28. $(x^{5n} + x^{3n} + 2)(x^{3n} + x^{2n} + x)$

Summary

1. The product of two terms with like signs is positive.
2. The product of two terms with unlike signs is negative.
3. The arithmetic coefficient of the product is the product of the absolute values of the coefficients of the terms being multiplied.
4. If more than two factors are multiplied together, the sign of the product is found by repeated multiplication of the negative or positive coefficients, expressed or implied.
5. To multiply powers of like bases, add the exponents over the base.
6. To multiply a polynomial by a monomial algebraically, multiply each term in the polynomial by the monomial, observing the law of exponents, and write the proper sign before each term in the product.
7. To multiply a polynomials algebraically, calculate the product of the multiplicand and each term in the multiplier, and calculate the algebraic sum of any similar terms.

Chapter 8

Division of Monomials and Polynomials

 ## 8.1 Introduction

The general rules for algebraic division of monomials and polynomials are implicit in the operation of algebraic multiplication, since division is the inverse of multiplication. However, the implied rules concerning division of terms entailing exponents may not be clearly apparent; hence, we will proceed step-by-step in our analysis.

The numerical coefficient of the quotient is easily calculated by dividing the absolute value of the coefficient in the dividend by the absolute value of the coefficient in the divisor. To determine the sign of the quotient, observe that the same relations apply as in multiplication. Therefore,

$$\frac{+1}{-1} = -1$$

and

$$\frac{-1}{+1} = -1$$

and

$$\frac{+1}{+1} = +1$$

and

$$\frac{-1}{-1} = +1$$

We recall that if a numerical coefficient is not expressed following a plus sign or a minus sign, a numerical coefficient of 1 is implied. The foregoing principles illustrate the following rules for calculating the sign of the quotient.

1. *The quotient is positive if the dividend and divisor have like signs.*
2. *The quotient is negative if the dividend and divisor have unlike signs.*

8.2 Subtraction of Exponents

We recall that the product of two powers that have the same base is calculated by writing the base to a power that is equal to the sum of the exponents. Since division is the inverse of multiplication, it follows that the quotient is calculated by writing the base to a power that is equal to the difference of the exponents. Thus,

$$\frac{A^m}{A^n} = A^{m-n}$$

In turn, we may write a general rule for calculation of the exponent in a quotient.

3. *To calculate the exponent of a quotient, subtract the exponent of the base in the divisor from the exponent of the base in the dividend.*

This general rule is also valid when the exponent of the base in the divisor is the same as the exponent of the base in the dividend.

Example 1

$$\frac{A^2}{A^2} = A^{2-2} = A^0 = 1$$

The general rule for calculation of the exponent in a quotient is also valid when the exponent of the base in the denominator is greater than the exponent of the base in the numerator.

Example 2

(a) $\dfrac{X^2}{X^7} = X^{2-7} = X^{-5}$

(b) $\dfrac{1}{X^7} = X^{-7}$ or $\dfrac{1}{X^{-7}} = X^7$

Since the fraction in Example 2(b) is the reciprocal, we are led to the following general rule.

4. *Any number raised to a power is equal to the reciprocal of the number raised to the same power, with the sign of the exponent changed.*

It follows from the foregoing equations that a number or an algebraic factor may also be taken from the numerator of a fraction and placed in the denominator, provided the sign of its exponent is changed.

Example 3

$$xy^2z^3 = \frac{y^2z^3}{x^{-1}} = \frac{xz^3}{y^{-2}} = \frac{z^3}{x^{-1}y^{-2}} = \frac{1}{x^{-1}y^{-2}z^{-3}}$$

Let us consider a practical example of the reciprocal of resistance in electrical circuits. We have become familiar with the fact that a resistor has a resistance value. However, we shall perceive that a resistor also has a *conductance* value. Conductance is defined as the reciprocal of resistance. Resistance is measured in ohms, and conductance is measured in Siemen. Resistance is symbolized by R, and ohms are symbolized by Ω. Conductance is symbolized by G. Thus,

$$R = \frac{1}{G} \, \Omega$$

$$G = \frac{1}{R} \, \text{S}$$

It follows that Ohm's law may be written in the following forms:

$$I = \frac{E}{R} \quad \text{or} \quad I = EG$$

$$E = IR \quad \text{or} \quad E = \frac{I}{G}$$

Example 4   Find the current in a series circuit in which a 3.3k ohm resistor is connected across a 12 volt supply.

$$G = \frac{1}{R} \approx \frac{1}{3.3\ k\Omega} \approx 0.303\ mS$$

$$I = \frac{E}{R} = \frac{12}{3.3k} \approx 3.64\ mA$$

also

$$I = EG = 12 \times 0.303\ mS \approx 3.64\ mA$$

Exercises 8–1   Divide and express all answers with positive exponents:

1. $\dfrac{a^3}{a^2}$

2. $\dfrac{x^5}{x^3}$

3. $\dfrac{3R^2}{R^2}$

4. $\dfrac{A^{3/2}}{A}$

5. $\dfrac{x^5}{x^9}$

6. $\dfrac{y^2}{y^3}$

7. $\dfrac{I^2}{I^7}$

8. $\dfrac{a^{3/2}}{a^{1/2}}$

9. $\dfrac{2^3}{2^2}$

10. $\dfrac{5^2}{5^3}$

11. $\dfrac{7^3}{7}$

12. $\dfrac{6^{1/2}}{6^{3/2}}$

13. $\dfrac{R^2}{R^{-3}}$

14. $\dfrac{E^3}{E^{-2}}$

15. $\dfrac{E^{-5}}{E^2}$

16. $\dfrac{C^{-3/2}}{C^{5/2}}$

17. $\dfrac{b^2}{b^{-3}}$

18. $\dfrac{1}{R^{-3}}$

19. A^{-5}

20. $\dfrac{10^{-3}}{10^2}$

21. $\dfrac{2^3}{2^{-3}}$

22. $\dfrac{3^{3/2}}{3^{-1/2}}$

23. $\dfrac{I^2R}{I^2}$

24. $\dfrac{2E^2}{2E}$

25. $\dfrac{\sqrt{3}}{3^{-2}}$

26. $\dfrac{\sqrt[3]{4}}{4^{-2/3}}$

27. $\dfrac{\sqrt{8^3}}{8}$

28. $\dfrac{\sqrt{3^3}}{3^{1/2}}$

29. $\dfrac{1}{2^{-3}}$

30. 4^{-2}

▶ 8.3   Division of Monomials

The foregoing discussion has introduced us to the algebraic division of monomials, and we will develop this topic in greater detail. Let us consider an example of division that involves concrete literal numbers. Figure 8-1 depicts a series-parallel resistive circuit. The parallel-connected resistors R_p have equal values, and the series resistor R_s has a value that is not the same as

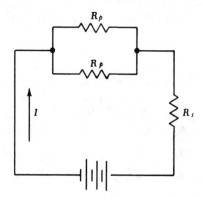

Figure 8–1
Series-parallel circuit.

R_p. The current value is denoted by I. If the power dissipated by one of the parallel-connected resistors has a value of P_p, and the power dissipated by the series resistor has a value of P_3, what is the ratio of total power dissipation by the parallel-connected resistors to the power dissipation by the series resistor?

We will observe that since the parallel-connected resistors have equal resistance values, half of the total current I, flows through each resistor, or the power dissipated in one resistor is formulated.

$$P_p = \left(\frac{I^2}{2}\right) R_p$$

In turn, it follows that the total power dissipation by the parallel-connected resistors is equal to $2P_p$.

Example 5

$$2P_p = \frac{2I^2 R_p}{4} = \frac{I^2 R_p}{2}$$

Next, the power dissipated by the series resistor is formulated.

$$P_1 = I^2 R_1$$

We write the ratio of $2P_p$ to P_3 as follows:

$$\frac{2P_p}{P_3} = \frac{I^2 R_p / 2}{I^2 R_3}$$

It is convenient to write the above formula in the form

$$\frac{2P_p}{P_3} = \frac{I^2 R_p}{2I^2 R_3}$$

The numerator is a monomial, and the denominator is a monomial. We obtain the answer by calculating their quotient.

$$\frac{2P_p}{P_3} = \frac{R_p}{2R_3}$$

Accordingly, we conclude that the ratio of total power dissipation by the parallel-connected resistors to the power dissipation by the series resistor in Figure 8-1 is equal to one-half of the ratio of R_p to R_3.

The foregoing demonstrations lead us to the general rule for algebraic divisions of monomials.

5. *To divide one monomial by another monomial, divide the coefficient of the dividend by the coefficient of the divisor, prefix the proper sign, and apply the rule of exponents to the literal factors.*

Exercises 8-2 Calculate the following quotients and express with positive exponents:

1. $25a^2 \div 5a$
2. $-12E^3 \div 8E$
3. $2\pi R^3 \div 4R^2$
4. $18E^2 P \div 6EP^2$

5. $2I_1^3 R_1 \div 4I_1^2 R_1$
6. $I^2 R \div IR$
7. $x^a \div x^b$
8. $\pi r^2 h \div 2\pi r$

9. $\frac{1}{3}\pi r^2 h \div \pi r^2 h$
10. $\frac{1}{2}gt^2 \div 2gt$
11. $lw \div 2\pi l w^2$
12. $x^{a+b} y^{a-b} \div x^{a-b} y^{a+b}$

13. $\sqrt{a} \div a^2$ **16.** $2gt^3 \div 3t^2$ **19.** $2N_1^2 \div N_1^3$

14. $(a^3)^2 \div (a^2)^3$ **17.** $4\sqrt{xy} \div 2x^2y^3$ **20.** $6a^5b^2c^3 \div 2a^2b^5c^3$

15. $a^{1/2}b^{3/2} \div a^{3/2}b^{1/2}$ **18.** $I^2E \div IR$

▶ 8.4 Division of a Polynomial by a Monomial

As would be anticipated, the algebraic division of a polynomial by a monomial is merely an extension of the rules for division of monomials. Let us consider an example that involves an electrical circuit. Figure 8-2 depicts a series-parallel circuit in which we will calculate the ratio of the power dissipated in all the resistors except R_3 to the power dissipated in R_3. The power dissipated in R_3 is evidently equal to $I^2R_2/3$. We wish to calculate the ratio of the power dissipated in all the resistors except R_3 to the power dissipated in R_3.

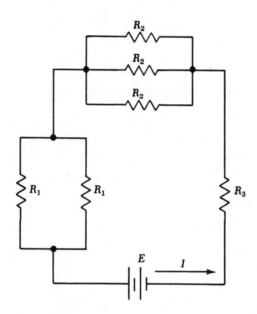

Figure 8–2
A series-parallel circuit in which we will formulate a power dissipation ratio.

Example 6   Therefore, we will write

$$\frac{(I^2R_1/2) + (I^2R_2/3)}{I^2R_3} = \frac{3I^2R_1 + 2I^2R_2}{6I^2R_3}$$

The numerator of the fraction is a binomial, which is to be divided by the denominator. The denominator is a monomial, which will be divided into each term of the numerator.

$$\frac{3I^2R_1 + 2I^2R_3}{6I^2R_3} = \frac{R_1}{2R_3} + \frac{R_2}{3R_3}$$

The foregoing examples lead us to the general rules for algebraic division of a polynomial by a monomial.

6. *Divide each term of the polynomial by the monomial, affix the sign of each term in the quotient by the rules for calculating the sign of a quotient; divide each literal factor in the polynomial by applying the rule for exponents in a quotient.*

Let us apply these general rules to another problem in which one of the power values is negative. With reference to Figure 8-2 we will calculate the ratio of the power dissipated in the two parallel combinations to the power supplied by the battery. Recall that the resistors are a *power sink*, and that the battery is a *power source*. Therefore, power values dissipated by the resistors are positive, and the power value supplied by the battery is negative. Accordingly, we write

Example 7

$$\frac{3I^2R_1 + 2I^2R_2}{-6EI} = \frac{3I^2R_1}{-6EI} + \frac{2I^2R_2}{-6E} = \frac{IR_1}{2E} - \frac{IR_2}{3E}$$

We have affixed a minus sign to each term in the quotient in accordance with Rule 6. If we wish to write a plus sign before a term in the quotient, we must then transfer the minus sign to either the numerator or the denominator.

$$-\frac{IR_1}{2E} - \frac{IR_2}{3E} = \frac{-IR_1}{2E} + \frac{-IR_2}{3E} = \frac{IR_1}{-2E} + \frac{IR_2}{-3E}$$

Exercises 8–3 Simplify the following expressions, and show the results in terms with positive exponents:

1. $(a^2x^2 + ax) \div a$
2. $(2a^2x^2 + 3ax) \div 2ax$
3. $(3x^3 + 6x^2 + 9x) \div (6x)$
4. $(E^2R + I^2R) \div IR$
5. $(2\pi r^2 + \pi r^3) \div \pi r^2$
6. $(65S^2 + 15S) \div 5S$
7. $(2a^2b^3c - 4ab^2c^3) \div 4ab^2c$

8. $(20a^3b^2c^2 + 12a^2bc^3) \div (-10a^2b^2c)$
9. $(2\pi h^3r + \frac{1}{3}\pi h^2r^2) \div 2\pi hr^2$
10. $(\frac{1}{2}gt^2 + = 2g^2t^3) \div 4gt^{-2}$
11. $(\frac{4}{3}\pi ab^2 - \frac{1}{2}\pi ab) \div 2\pi a^2b^3$
12. $(\frac{\pi}{3}hr_1^2 + \frac{\pi}{3}hr_1r_2 + \frac{\pi}{3}hr_2^2) \div \frac{\pi}{3}hr_1r_2$
13. $(\pi r^3 + \pi rh^2) \div \frac{1}{6}\pi hr^2$
14. $(I_1^2R_1^2R_2 - I_2^2R_1R_2^2) \div II_2R_1$

▶ 8.5 Division of a Polynomial by Another Polynomial

The algebraic division of a polynomial by another polynomial can be compared with long division of whole numbers, just as the previous examples of algebraic division can be compared with short division of whole numbers. Let us take an example of the division of polynomials with the following dividend and divisor:

Example 8

$$\frac{a^2x + a^3 - x^3 - ax^2}{a + x}$$

It is convenient to arrange the dividend and divisor in descending powers of the same letter. Thus, let us write the fraction in the form

$$a^3 + a^2x - ax^2 - x^3$$

Divide

$$
\begin{array}{r}
a^2 \qquad\quad - x^2 \\
a + x\,\overline{\big)\;a^3 + a^2x - ax^2 - x^3} \\
a^3 + a^2x \qquad\qquad \\
\overline{\;0 + 0 \quad - ax^2 - x^3} \\
\overline{0 \quad + 0}
\end{array}
$$

Observe that we divided the first term in the dividend, a^3, by the first term in the divisor, a. In turn, we wrote a^2 as the first term in the quotient. Then, we multiplied the entire divisor, $a + x$, by the first term in the quotient, a^2. The product was $a^3 + a^2x$, which we wrote under corresponding terms in the dividend. Then we subtracted, and obtained a difference of zero. Accordingly, we brought down the next two terms from the dividend, $-ax^2 - x^3$. We calculated the second term in the quotient by dividing $-ax^2$ by the first term in the divisor, a. This gave us $-x^2$ for the second term in the quotient. Finally, we multiplied the entire divisor, $a + x$, by $-x^2$, and wrote the product, $-ax^2 - x^3$ under the corresponding terms, $-ax^2 - x^3$.

The quotient is $a^2 - x^2$, and there is no remainder in the foregoing example. Of course, we will often encounter problems in which there is a remainder.

Example 9

$$
\begin{array}{r}
2R^3 + 3R^2 + 3R + 3 \\
3R - 2\,\overline{\big)\;6R^4 + 5R^3 + 3R^2 + 3R + 2} \\
6R^4 - 4R^3 \qquad\qquad\qquad\quad \\
\overline{\;9R^3 + 3R^2 \qquad\qquad\quad} \\
9R^3 - 6R^2 \qquad\qquad\quad \\
\overline{\;9R^2 + 3R \qquad\quad} \\
9R^2 - 6R \qquad\quad \\
\overline{\;9R + 2} \\
9R - 6 \\
\overline{\;8 \ (\text{Remainder})}
\end{array}
$$

The quotient is $2R^3 + 3R^2 + 3R + 3$, with a remainder of 8. As in the familiar case of long division with whole numbers, we write the quotient as follows:

$$
2R^3 + 3R^2 + 3R + 3 + \frac{8}{3R - 2}
$$

These examples lead us to the general rules for the division of one polynomial by another polynomial.

7. *Choose a common literal factor, and rearrange the dividend and divisor (if necessary) in descending powers of this literal factor.*
8. *Calculate the first term in the quotient by dividing the first term in the dividend by the first term in the divisor.*
9. *Calculate the product of the first term in the quotient and the entire divisor; write this product under corresponding terms in the dividend and subtract this product from the corresponding terms of the dividend.*
10. *Bring down the next term or terms in the dividend as in ordinary long division, and repeat the foregoing operations until there is a remainder of zero, or a remainder that cannot be further divided by the divisor.*

11. *If there is a remainder that cannot be further divided by the divisor, write it as the numerator of a fraction, with the divisor as denominator, and write this fraction as the final term in the quotient.*

Exercises 8–4 Perform the following operations:

1. $(3a^2 + 8a + 9) \div a + 2$
2. $(a^2 - 4b^2) \div (a - 2b)$
3. $(x^2 + 11x + 24) \div (x + 3)$
4. $(y^5 - 1) \div (y - 1)$
5. $(x^2 - y^2) \div (x - y)$
6. $(x^3 - y^3) \div (x - y)$
7. $(x^3 - y^3) \div (x^2 + xy + y^2)$
8. $(625x^2 - 81) \div (5x - 3)$
9. $(12E^4 - 3ER^3 - 24E^3R + 15E^2R^2) \div (6E^2 - 3ER)$
10. $(x^2 + xy + y^2) \div (x - y)$
11. $(x^2 - xy + y^2) \div (x + y)$
12. $(2E^2 + 3E - 1) \div (2E^2 - 3E + 1)$
13. $(30R^4 + 11R^3 - 82R^2 - 5R + 3) \div 30R^2 + 2R - 4$
14. $(x^2 - y^2 + 2xz - z^2) \div (x + y + z)$
15. $(8I - 13I^2 + 12) \div (2I - 3)$

Summary

1. *Division*—In division of one term by another, the coefficient, the sign, and the exponent must be considered in obtaining the quotient.
2. *The Sign*—The quotient of quantities with like signs is positive; the quotient of quantities with unlike signs is negative.
3. *The Exponent*—The quotient of terms having like bases has the same base as the terms and an exponent which is the difference of the exponents of the terms.
4. *The Coefficient of the Quotient*—is obtained by dividing the absolute value of the coefficient of the dividend by the absolute value of the coefficient of the divisor.
5. *Quantities with Zero as an Exponent*—any quantity with an exponent of zero is equal to one.
6. *Quantities with Negative Exponents*—any quantity with a negative exponent is equal to the reciprocal of that quantity with a positive exponent. Thus,
7. *Moving Factors to the Numerator or Denominator*—any factor may be moved from the numerator to the denominator or from the denominator to the numerator by changing the sign of the exponents.

Chapter 9

Equations and Formulas

 ## 9.1 Introduction

By definition, the *solution* of an equation or formula entails calculation of the value or values of the unknown number or numbers that will *satisfy* the equation or formula. To *satisfy* an equation or formula is to substitute values that make the equation or formula an identify.

An equation is satisfied if it is a mathematical identity in terms of abstract numbers. A formula is satisfied if it is made a mathematical and physical-unit identity in terms of concrete numbers. We call this value or these solutions the *root* or *roots* of the equation or formula.

 ## 9.2 Axioms for the Solution of Equations

An *axiom* is a self-evident truth or accepted principle that cannot be proven. For example, by definition, lines that lie in the same plane but do not meet, however far they may be extended, are said to be parallel. To most people, it is self-evident that if the lines are drawn the same distance apart at both ends, they will not meet, however far they may be extended. Because no one can *prove* that the lines will never meet, we usually accept the foregoing as an axiom.

Various other axioms are necessary to solve equations and formulas. Let us consider some of these axioms.

1. *If the same number is added to both members of an equation, the new equation is true.*

Example 1

$$Y = Y$$

Add 100 to each side.

$$Y + 100 = Y + 100$$

If we substitute 1000 for Y in the foregoing equations, we obtain

$$Y = Y$$
$$1000 = 1000$$
$$Y + 100 = Y + 100$$
$$1100 = 1100$$

2. *If the same number is subtracted from both members of an equation, the new equation is true.*

Example 2   In the equation $x = x$, let $x = 10$.

$$10 = 10$$

Subtract 2 from each member.

$$x - 2 = x - 2$$
$$10 - 2 = 10 - 2$$

Then

$$8 = 8$$

3. *If both members of an equation are multiplied by the same number, the new equation is true.*

Example 3 Multiply the equation $x - 3 = 5$ by 10.

$$(x - 3)10 = 5 \times 10$$
$$10x - 30 = 50$$

Solution:

$$10x = 80$$
$$x = 8$$

Proof by substitution:

$$(8 - 3)10 = 5 \times 10$$
$$50 = 50$$

4. *Both members of an equation may be divided by the same number without destroying the equality.*

Example 4

$$x = y$$

For any value of x, y is the same value; let $x = 20$.

$$20 = 20$$

then

$$\frac{x}{5} = \frac{y}{5}$$

and

$$\frac{20}{5} = \frac{20}{5}$$
$$4 = 4 \qquad \text{(Answer)}$$

5. *Numbers that are equal to the same number are equal to each other.*

Example 5

$$R = a \text{ and } R = b$$

then

$$a = b$$

because

$$R = 100$$

$$a = 100 \quad \text{and} \quad b = 100 \qquad \text{(Answer)}$$

6. *Like powers of equal numbers are equal.*

Example 6 If

$$R = R$$

then

$$R^4 = R^4$$

because, if $R = 2$,

$$2^4 = 2^4$$

and

$$16 = 16$$

7. *Like roots of equal numbers are equal.*

Example 7 If

$$R = R$$

then

$$\sqrt{R} = \sqrt{R}$$

because, if $R = 100$,

$$\sqrt{100} = \sqrt{100}$$

and

$$10 = 10$$

▶ *9.3   Rules for the Solution of Equations*

The foregoing axioms lead to the statement of important rules for use in the solution of equations and formulas. Demonstrations of the following rules are given in the form of examples.

8. *Any term can be removed from one member of an equation and placed in the other member, provided the sign of the term is changed.*

Example 8   If

$$x - 5 = 10$$
$$x = 10 + 5 = 15$$

Checking by substitution,

$$15 - 5 = 10$$
$$10 = 10 \qquad \text{(Answer)}$$

   9. *Any term that appears in both members of an equation can be canceled from both members, provided that the sign of the term is the same in each member.*

Example 9   If

$$a + b = c + b$$
$$a = c + b - b$$
$$a = c \qquad \text{(Answer)}$$

  10. *All of the signs in any equation can be changed without destroying the identity.*

Example 10   If

$$a - 3 = 9$$
$$a = 9 + 3$$
$$a = 12$$

or

$$-a + 3 = -9$$
$$-a = -9 - 3$$

and multiplying each side by (-1),

$$a = 12 \qquad \text{(Answer)}$$

Exercises 9–1   Solve the following problems for the unknown, and check your answers:

1.	$x - 2 = 3$	**13.**	$6R = 24$
2.	$2y + 3 = 5$	**14.**	$6R = 3(R + 1)$
3.	$2a + 3 = a + 2$	**15.**	$-R = -(15 + 10)$
4.	$R - 15 = 3R + 5$	**16.**	$IR_1 = I(2 + 6)$
5.	$2I_1 - 3 = I_1 - 7$	**17.**	$2ab = a(b + 4)$
6.	$m - 6 + 6 = 12 - 5$	**18.**	$R^2 = 25$
7.	$a + 4 + 5 = -a + 2 + 3$	**19.**	$(2R)^2 = (R + 3)^2$
8.	$2r + \frac{1}{2} = r + \frac{1}{4}$	**20.**	$(2R)^{1/2} = (R - 3)^{1/2}$
9.	$8q + q = -2 + 21$	**21.**	$5a + 3 + 2a = 3a + 5 - 6$
10.	$2R + 3R - 6 = 24$	**22.**	$25 - (i + 4) = 10 - (2i + 3)$
11.	$3B + 3 = B + 2$	**23.**	$5P - 2(P - 2) = P + 3$
12.	$5x + 10 = x + 50$	**24.**	$7(r + 2) - 3(2 - r) = 16 + 3(2 + r)$

25. $2 + 5(e - 5) - 2(2e + 3) = 0$

26. $5(m + 1) - 2(3m - 3) = m^2 - (m^2 + 2)$

27. $n(2n + 1) - 2(n^2 + 3) = 2n + 6$

28. $b^2 + 3b - 2(b - 3) - b(b - 6) = 0$

▶ 9.4 *Mathematical Statements of Circuit Problems*

Mathematics is a language that makes *exact* statements about the relations among *numbers*. On the other hand, the English language makes descriptive statements about physical *objects* and physical *units*, and these descriptive statements are inexact. For example, the mathematical statement "$2 + 2 = 4$" is an exact statement about additive relations among the numbers 2, 2, and 4. However, the English statement, "We calculate the value of resistance in a lamp filament from voltmeter and ammeter readings," is a descriptive and inexact statement about physical objects: a lamp filament, a voltmeter, and an ammeter. Note that it is a descriptive and inexact statement about the relations of *measured values of physical units:* volts, amperes, and ohms.

To describe the circuit in Figure 9-1 mathematically, we must first compile pertinent statements in English concerning the physical units that can be assigned. Then, we must make measurements of the values of certain physical units. The value of one physical unit is unknown. Our problem is to calculate the value of the unmeasured physical unit. We proceed as follows.

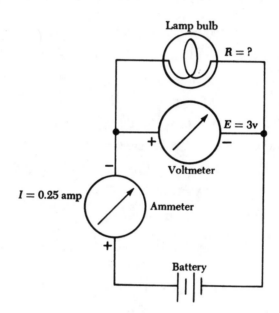

Figure 9–1
Statements in English concerning this circuit are translated into a literal algebraic formula.

First, we *translate* the English statements into literal algebraic formula. Second, we *substitute* the measured values into this formula. Third, we *solve* the formula for the unknown value. Let us list the observations and steps we take to calculate the resistance of the lamp filament in Figure 9-1.

1. The circuit comprises a lamp connected in series with a battery.
2. A voltmeter indicates that 3 V are being applied to the filament.
3. An ammeter indicates that 0.25 amp is flowing through the lamp filament.
4. Question: What is the resistance value of the lamp filament?
5. Translate the English statements number 1, 4, and 5 into an algebraic formula yields

$$R = \frac{E}{I}$$

6. Substitution of measured values number 2 and 3 into the formula provides

$$R = 3/0.25 \ \Omega$$

7. Solution of this formula results in the answer

$$R = 12 \ \Omega$$

Exercises 9–2 Substitute the values given into these formulas, and calculate the result:

1. $A = LW$; $L = 25$ ft, $W = 12$ ft
2. $S = \frac{1}{2}gt^2$; $g = 32, t = 5$ sec
3. $R = \frac{E}{I}$; $E = 25$ v, $I = 200$ amp
4. $A = \pi r^2$; $r = 6$ cm
5. $A = \frac{1}{2}bh$; $b = 20$ ft, $h = 10$ ft
6. $C = \frac{5}{9}(F - 32)$; $F = 52°$ F
7. $f = \frac{1}{T}$; $T = 2 \ \mu$sec
8. $x_L = 2\pi fL$; $f = 100$Hz, $L = 0.001$h
9. $V = LWH$; $L = 2$ cm, $W = 5$ cm, $H = 18$ cm
10. $V = \frac{1}{3}\pi r^2 h$; $r = 2$ in., $h = 15$ in.
11. $Q = CE$: $C = 100 \ \mu$f, $E = 200$ v
12. $M = K\sqrt{L_1 L_2}$; $K = 1, L_1 = 100 \ \mu$h; $L_2 = 25 \ \mu$h
13. $e = E - IR$; $E = 100$ v, $I = 2$ amp, $R = 52 \ \Omega$
14. $\mu = G_m r_p$; $G_m = 100 \ \mu\mho, r_p = 10 \ k\Omega$
15. $pf = \frac{R}{X}$; $r = 100 \ \Omega, X = 200 \ \Omega$

Solve the following formulas for the letter indicated:

	Formula:	Solve for:	Description of Formula:
1.	$I = E/R$	R	Ohm's law
2.	$P = I^2 R$	I	Power
3.	$Z^2 = R^2 + X^2$	R	Impedance
4.	$X_L = 2\pi fL$	f	Inductive reactance
5.	$Q = CE$	C	Coulomb charge
6.	$\mu = g_m r_p$	r_p	Vacuum tube gain
7.	$\frac{N_p^2}{N_P^2} = \frac{Z_p}{Z_s}$	N_p	Transformer
8.	$\phi = HA$	A	Flux density
9.	$pf = \frac{R}{X}$	X	Power factor
10.	$f = \frac{Q_1 Q_2}{Kd^2}$	Q_1	Coulomb's electrostatic law

Formula:	Solve for:	Description of Formula:
11. $R_t = R_0(1 - \delta t)$	R_0	Resistance of metal
12. $M = KIt$	I	Faraday's law of electrolysis
13. $R = \rho \dfrac{L}{A}$	A	Resistance of a conductor
14. $R_{\text{int}} = \dfrac{E_\alpha - E_{cc}}{I}$	E_α	Internal battery resistance
15. $H = \dfrac{NI}{L}$	I	Magnetic field intensity
16. $R_T = R_1 + R_2 + R_3$	R_3	Resistance of a series circuit
17. $G_T = G_1 + G_2 + G_3$	G_2	Conductance of a parallel circuit
18. $\beta = \dfrac{I_C}{I_B}$	I_B	Gain of a transistor
19. $Q = \dfrac{X_L}{R}$	R	Q of a coil
20. $E_s = j\omega m i_p$	i_p	Voltage in a tuned transformer

Summary

1. An equation is defined as a mathematical statement that two quantities are equal.
2. A conditional equation has members that are equal for a limited number of numerical values.
3. If the same quantity is added to both sides of an equation, the new equation is true.
4. If the same quantity is subtracted from both members of an equation, the new equation is true.
5. If both sides of an equation are multiplied by the same quantity, the new equation is true.
6. If both sides of an equation are divided by the same quantity, the new equation is true.
7. Numbers that are equal to the same number are equal to each other.
8. Like powers of equal numbers are equal.
9. Like roots of equal numbers are equal.
10. Any term can be removed from one member of an equation and placed in the other member, provided the sign of the term is changed.
11. Any term that appears in both members of an equation can be canceled from both members, provided that the sign of the term is the same in each member.
12. All the signs of an equation can be changed without destroying the identity.

Chapter 10

Factors of Algebraic Expressions

 ## 10.1 Introduction

We have become familiar with the factors of simple algebraic expressions in earlier work. At this point, we will define algebraic factors as *rational integral expressions that divide an algebraic expression without a remainder*. Therefore, an algebraic expression may be a single term, a monomial, or a polynomial.

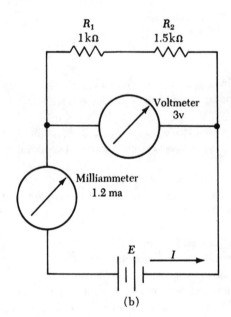

Figure 10–1
Circuit with two resistors, R_1 and R_2 in series, through which current I flows.

 ## 10.2 Powers and Roots of Monomials

The operation of *factoring* an algebraic expression denotes the calculation of two or more expressions whose product is equal to the original algebraic expression. In simple situations, factoring may be accomplished by inspection.

Example 1   The power dissipated as heat by the resistors in the circuit if Figure 10-1 is formulated

$$I^2R_1 + I^2R_2 = I^2(R_1 + R_2)$$

It is evident by inspection of the formula in Example 1 that I^2 and $(R_1 + R_2)$ are the factors of $I^2R_1 + I^2R_2$.

Example 2   We recognize that the power P_1 dissipated as heat in resistor R_1 may be formulated

$$(E - IR_2) = E_1$$

and

$$\frac{E - IR_1}{R_2} = \frac{E_2}{R_2} = I_t$$

$$\therefore (E - IR_2)\left(\frac{E - IR_1}{R_2}\right) = E_1 I_t = P_1$$

Let us review briefly the rules for calculation of powers of monomials and for extraction of roots of monomials.

1. *The square of a monomial is equal to the square of the numerical coefficient, multiplied by the product of the literal factors, with the exponent of each literal factor multiplied by 2.*

Example 3

$$y = (3a - 2b)^2 = (3a - 2b)(3a - 2b) = 9a^2 - 12ab + 4b^2$$

Next, we state the general rule for calculation of the cube of a monomial.

2. *The cube of a monomial is equal to the cube of the numerical coefficient, multiplied by the product of the literal factors, with the exponent of each literal factor multiplied by 3, and the sign of the answer is the same as the sign of the monomial.*

Example 4

$$(xy^2z^2)^3 = x^3y^6z^6$$
$$(-2xy^2)^3 = -8x^3y^6$$

Next, let us review briefly the extraction of a square root. We may state the following general rule.

3. *The square root of a monomial is equal to the square root of the numerical coefficient, multiplied by the literal factors, with the exponent of each literal factor divided by 2, and the sign of the answer is $+$ or $-$.*

Example 5

$$\sqrt{9x^4y^6} = \pm 3x^2y^3$$

Every abstract number has two square roots which are equal in absolute value and opposite in sign. We call the positive root the *principal root*.

When we work with formulas only the principal root might correspond to a real physical relation. Thus, if the area of a square is 9 sq in., we can accept the root $+3$ in. only, and we must reject the root -3 in. for the length of a side of the square, because we cannot draw a square that has a side of length less than zero.

Next, let us review briefly the extraction of a cube root. We may state the following general rule.

4. *The cube root of a monomial is equal to the cube root of the numerical coefficient, multiplied by the literal factors, with the exponent of each literal factor divided by 3, and the sign of the answer is the same as the sign of the monomial.*

Example 6

$$\sqrt[3]{27x^3y^6z^6} = 3xy^2z^2$$

$$\sqrt[3]{-27x^3y^6z^6} = -3xy^2z^2$$

but

$$\sqrt[-3]{-27x^3y^6z^6} = 3xy^2z^2$$

Exercises 10–1   Factor the following:

1. $ax^2 + x$

2. $2a^2x^2 + ax^3$

3. $2IR^2 + 4I^2R$

4. $2ab^2c + a^2b^3c^2$

5. $4x^2y^3 + 2xy^2$

6. $3e^2 + 12e^2c$

7. $2\frac{a^2}{b} - 4\frac{a}{b^2}$

8. $5f^2g^3 - 25f^3g^4$

Find the values of the indicated power:

9. (πr^2)

10. $(3\pi h)^2$

11. $(abc^2)^2$

12. $(3a^2b^3)^2$

13. $(3x^2y)^2$

14. $(2a^{-1})^2$

15. $(5IR^{-2})^2$

16. $-(5x^2yz^{-1})^2$

17. $(-x^2y^2z^3)^2$

18. $(-3x^2y)$

19. $(-5I^{-2})^2$

20. $(-a^3b^2c^{-2})^2$

21. $\left(\dfrac{2\pi}{3T}\right)^2$

22. $\left(\dfrac{-2\pi}{3}\right)^2$

23. $-\left(\dfrac{2\pi^2}{5E}\right)^2$

24. $-\left(\dfrac{-2a^2b^3}{R}\right)^2$

25. $(2\pi)^3$

26. $(-3\pi r)^3$

27. $-(2\pi r)^3$

28. $(ab^3y)^3$

29. $(ap^{-2})^3$

30. $(-3x^2yz^{-2})^3$

31. $(-2x^3y)^3$

32. $-(2x^2y)^3$

33. $\left(\dfrac{2h^2}{3}\right)^3$

34. $\left(\dfrac{2h^{1/2}}{r^2}\right)^3$

35. $\left(-\dfrac{2c^{1/3}}{3}\right)^3$

36. $-\left(\dfrac{ax^2}{-by}\right)^3$

37. $-\left(\dfrac{-iR^2}{T}\right)^3$

38. $\left(\dfrac{-aT^{-2}}{4}\right)^2$

Find the value of the indicated roots:

39. $\sqrt{I^2R^2}$

40. $\sqrt{ab^2}$

41. $\sqrt{4\pi^2h^2}$

42. $-\sqrt{36a^4R^6}$

43. $\sqrt{ab^2c}$

44. $\sqrt{\dfrac{9R^2}{y^6}}$

45. $-\sqrt{\dfrac{\pi 2}{R^4}}$

46. $\sqrt{\dfrac{1}{9}a^2b^4}$

47. $\sqrt{\dfrac{4a^2b^2}{x^4y^6}}$

48. $\sqrt{\dfrac{9R^2I^2}{25p^4}}$

49. $\sqrt{\dfrac{36R^6}{25p^2}}$

50. $\sqrt{\dfrac{25a^2b^{-2}}{c^6}}$

51. $\sqrt{\dfrac{1}{16}x^6y^8z^{-2}}$

52. $\sqrt[3]{8}$

53. $\sqrt[3]{-27}$

54. $-\sqrt[3]{-81}$

55. $\sqrt[3]{\dfrac{27}{8}}$

56. $\sqrt[3]{\dfrac{x^5}{x^2}}$

57. $\sqrt[3]{-\dfrac{1}{8}}$

58. $\sqrt[3]{\dfrac{27}{125}}$

59. $\sqrt[3]{8x^3y^9}$ **60.** $\sqrt[3]{\dfrac{ax^3y^2}{8}}$ **61.** $\sqrt[3]{27x^3y^6}$ **62.** $\sqrt[3]{125x^3y^{-8}}$

63. $\sqrt[3]{48R^4I^6}$ **64.** $\sqrt[3]{-x^5y^{-2}}$ **65.** $\sqrt[3]{\dfrac{3375}{r^3}}$ **66.** $\sqrt[3]{-\dfrac{M^6}{8N^3}}$

67. $\sqrt[3]{\dfrac{729R^3I^6}{8M^3N^3}}$ **68.** $-\sqrt[3]{\dfrac{\beta^6r^3}{\gamma^6\alpha^3}}$ **69.** $\sqrt[3]{\dfrac{\pi^{1/3}r^{1/2}}{\beta^{1/2}}}$ **70.** $\sqrt[3]{\dfrac{1/2}{\pi^{1/2}t^{1/4}}}$

Problems 10–1

1. Find the power dissipated as heat in the circuit in Figure 10-1; $E_1 = 100\,\Omega$, $R_2 = 230\,\Omega$, and $I_T = 200$ ma.
2. What is the current flow through an 82-ohm resistor in which 1,024 w are dissipated?
3. What is the volume of a cylinder with a radius of 20 cm and a height of 150 cm? $(V = 2\pi r^2 h)$
4. What is the volume of a spheroid with minor axis $a = 2$ in. and major axis $b = 8$ in.? $(V = 4\pi a^2 b/3)$
5. What is the radius of a sphere with a volume of 256 cm³? $(V = 4\pi r^3/3)$
6. Find the radius of a sphere with a surface area of 1,600 cm². $(S = \pi d^2)$
7. Find the angle θ in degrees between the radii of a circle sector with a radius of 2 in. and an area of 0.528 in.² $(A = \pi r^2 \theta/360)$
8. What is the stored energy in watt-seconds in a 100-microfarad capacitor with 100 v across the plates? $(W = CE^2/2)$
9. Find the current flow through an inductor of 10 mh which causes an energy storage of 1,000 wsec. $(W = LI^2/2)$
10. What is the magnetic pull force in pounds of a magnet with an area (A) of 1 sq in. and a flux density (B) of 40, 000 lines per in.²? $(F = B^2A/7213 \times 10^4)$

 10.3   Monomial Factors of Polynomials

Previous discussion has introduced the concept of a monomial factor of a polynomial. Let us consider a practical example, with reference to a series circuit containing three resistors. The current flow in the circuit is stated by Ohm's law, $I = E/R_T$, where $R_T = R_1 + R_2 + R_3$. In turn, we may formulate the sum of the resistive voltage drops as follows:

$$E = V_1 + V_2 + V_3$$

$$I_T R_T = I_T R_1 + I_T R_2 + I_T R_3$$

$$\frac{E}{R_T}R_T = \frac{E}{R_T}R_1 + \frac{E}{R_T}R_2 + \frac{E}{R_T}R_3$$

The left-hand member of the formula contains E as a common monomial factor, and also contains $1/R_T$ as a common monomial factor. The greatest common factor is E/R_T. Note that the right-hand member of the formula represents the result of factoring the left-hand member. This is an example of the following general rule.

5. *To factor a polynomial that has terms containing one or more common monomial factors, inspect the polynomial for its greatest common factor, calculate the quotient of the polynomial divided by this greatest common factor, and write the answer with the quotient enclosed in parentheses preceded by this greatest common factor.*

Example 7 We observe that the following polynomial has two greatest common factors.

$$ax^2y - bxyz - m^2n - m^3p = xy(ax - bz) - m^2(n + mp)$$

Algebraic signs must be carefully observed in factoring operations. When concrete literal numbers are used in formulas, the sign of the number is originally the same as the sign of the physical unit. For example, a circuit might entail current flows in opposite directions through a resistor. If I_A is designated as a positive current, then I_B must be designated as a negative current with respect to the resistor.

 ## 10.4   Calculation of the Square of a Binomial

To develop the general rule for calculation of the square of a binomial, let us observe a specific example, and analyze the algebraic relations among the terms of the binomial and the terms of the product. With reference to Figure 10-1, the student may show that

$$\frac{E^2}{I^2} = (R_1 + R_2)^2$$

This formula follows from Ohm's law, when both members of the formula are squared. Let us calculate the square of $R_1 + R_2$ step-by-step.

$$
\begin{array}{ll}
\text{Multiplicand,} & R_1 + R_2 \\
\text{Multiplier,} & R_1 + R_2 \\
\hline
& R_1^2 + R_1R_2 \\
& \qquad R_1R_2 \ + R_2^2 \\
\hline
\text{Product,} & R_1^2 + 2R_1R_2 + R_2^2 \quad \text{(Answer)}
\end{array}
$$

$$(R_1 + R_2)^2 = R_1^2 + 2R_1R_2 + R_2^2$$

This formula is an example of the following general rule.

6. *The square of a binomial that is the sum of two positive terms is equal to the square of the first term, plus twice the product of the terms, plus the square of the second term.*

Let us apply this general rule to two other examples.

Example 8

$$(x^{-1} + y)^2 = x^{-2} + 2x^{-1}y + y^2$$

and

$$(\sqrt{2}ab^{-1} + \sqrt{3}c^2d)^2 = 2a^2b^{-2} + 2\sqrt{2}\sqrt{3}ab^{-1}c^2d + 3c^4d^2$$

Rule 6 is very useful, because it lessens the number of calculations that are required to square a binomial. Let us consider one more example.

Of course, we will often encounter binomials that are the difference of two positive terms or the sum of a positive term and a negative term. Consider, as a practical example,

Example 9 If the voltage drop across resistor R_1 in Figure 10-1 is symbolized V_1, the student may show that

$$E_1^2 = (E - IR_2)^2$$

Let us calculate the square of $E - IR_2$ step-by-step.

$$
\begin{array}{ll}
\text{Multiplicand,} & E - IR_2 \\
\text{Multiplier,} & \underline{E - IR_2} \\
 & E^2 - EIR_2 \\
 & \underline{\quad - EIR_2 + I^2R_2^2} \\
\text{Product,} & E^2 - 2EIR_2 + I^2R_2^2 \quad \text{(Answer)}
\end{array}
$$

This is an example of the following general rule.

7. *The square of a binomial that is the difference of two positive terms is equal to the square of the first term, minus twice the product of the terms, plus the square of the second term.*

Note that if we should be required to calculate the square of $-IR_2 + E$ it would be convenient to write the terms in reverse order, and square the binomial $E - IR_2$. An interesting situation is encountered when we square a binomial such as $-E - IR_2$. Let us calculate this square step-by-step.

$$
\begin{array}{ll}
\text{Multiplicand,} & -E - IR_2 \\
\text{Multiplier,} & \underline{-E - IR_2} \\
 & E^2 + EIR_2 \\
 & \underline{\quad EIR_2 + I^2R_2^2} \\
\text{Product,} & E^2 + 2EIR_2 + I^2R_2^2 \quad \text{(Answer)}
\end{array}
$$

Observe that the product we obtain is the same as when we square $E + IR_2$. Recall that $X^2 = (-X)^2$; accordingly, if we define $X = (E + IR_2)$, and $-X = (-E - IR_2)$, it becomes evident that $(E + IR_2)^2 = (-E - IR_2)^2$.

Exercises 10–2   Find the products:

1. $2a^2(x^2 + y^2)$

2. $I(R_1^2 + ER_2)$

3. $E^2\left(\frac{1}{R^2} + \frac{1}{R^3}\right)$

4. $(a + b)^2$

5. $(a - b)^2$

6. $(i - 2R)^2$

7. $(x + y)(x - y)$

8. $(\theta - \phi)(\theta + \phi)$

9. $(-a - b)(a + b)$

10. $(2R + I)(2R - I)$

11. $(E - 5)^2$

12. $(R^2 + 11)^2$

13. $\left(\frac{1}{2}R + 1\right)^2$

14. $(6R - 7I)^2$

15. $(a^3y^2 + b^2)^2$

16. $(3B + \theta)^2$

17. $(-RI_1 - RI_2)^2$

18. $(-E_1 + E_2)^2$

19. $(5R - IE)^2$

20. $(-R - I)^2$

21. $(2R + 3I)(2R - 3I)$

22. $(x^2 - y^2)(x^2 + y^2)$

23. $(-x^5 + 7)(-x^5 + 7)$

24. $(3I - 2)(3I - 5)$

25. $(2B + 7)(2B + 3)$

26. $(n^5 + 2)(n^5 - 3)$

27. $(0.1R - 0.5Z)(0.1R - 2Z)$

28. $(2Z + R)(2Z - R)$

Supply the missing term:

29. $x^2 + ? + 1$

30. $x^2 - ? + 4$

31. $R^2 + 6R + ?$

32. $25I^2 - 90I + ?$

33. $? + 14\theta + 49$

34. $9Z^2 - ? + 16$

35. $36Z^2 + ? + R^2$

36. $? - 12I + 9$

Factor:

37. $x^2 + 4x + 4$

38. $R^2 + 10R + 21$

39. $z^2 + 3z - 10$

40. $I^2 - 11I + 30$

41. $z^2 - 3z - 10$

42. $25\theta^2 + 5\theta - 12$

43. $x^2 - 13xy + 36y^2$

44. $2y^2 - y - 3$

45. $W^2 - \frac{1}{4}W + \frac{1}{64}$

46. $a^2 - a + \frac{1}{4}$

47. $4I^2R + 8IRE + R^2E$

48. $12y^2z - 6yz^2 - 18z^2$

▶ 10.5 The Product of Two Binomials

Instead of multiplying two binomials by the step-by-step process, we can save time by using the sequential diagram depicted in Figure 10-2. The procedure entails the calculation of four products added to one another.

Figure 10-2
Sequential diagram for multiplication of two binomials.

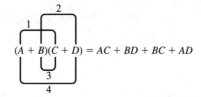

$$(A + B)(C + D) = AC + BD + BC + AD$$

When negative signs appear in the terms of the binomials, we must, of course, observe the algebraic laws of signs.

Exercises 10-3

1. $(a - b)(a + 2b)$

2. $(-x + y)(-x + 2y)$

3. $(2I - R)(I + 2R)$

4. $(x + 5)(x - 7)$

5. $(I - 3R)(I + 5R)$

6. $(Z + R)(-Z - 2R)$

7. $(2x + 5)(2x + 7)$

8. $(a + by)(a + cy)$

9. $(a + b)(a - 2b)$

10. $\left(\frac{1}{2}I + \frac{1}{4}R\right)\left(I - \frac{1}{2}R\right)$

11. $(n^5 + 11)(2n^2 - 5)$

12. $(2I + 3R)(2I + 7R)$

13. $(\Phi + \theta)(\Phi - \theta)$

14. $(\omega^2 - \lambda)(2\omega - \lambda)$

15. $(2t - 3T)(-t + 2T)$

16. $(L^2 - 3C^2)(2\pi L + C)$

17. $(0.1v - w)(0.2v + 0.3w)$

18. $(3e - 0.5)(2e + 0.3)$

19. $(1.5r - 2.1i)(0.6r + 1.6i)$

20. $\left(\frac{1}{8}\pi - 2\Phi\right)\left(\frac{1}{4}\pi - \frac{1}{2}\Phi\right)$

 ## 10.6   *Extraction of the Square Root of a Trinomial*

Since the trinomial $x^2 + 2xy + y^2$ is the exact, or perfect, square of $x + y$, and the trinomial $x^2 - 2xy + y^2$ is the perfect square of $x - y$, it is evident that the square root of a perfect trinomial square can be extracted by inspection. Moreover, since the trinomial $x^2 + 2xy + y^2$ is the perfect square of $-x - y$, it is obvious that the trinomial has two square roots, just as 4 has the square roots ± 2.

Thus, a trinomial is a perfect square provided two of its terms are squares of monomials and positive, and its other term is twice the product of the monomials; this term may have either a plus or a minus sign. These considerations lead us to the general rule for extraction of the square root of a perfect trinomial square.

8. *The square root of a perfect trinomial square is equal to the square roots of the two perfect monomial squares; if the sign of the other trinomial term is positive, the square roots have like signs; if the sign of the other trinomial term is negative, the square roots have unlike signs.*

 ## 10.7   *The Difference of Two Algebraic Numbers*

We will next consider the factors of the difference of two algebraic numbers, such as $x^2 - y^2$. It is helpful to observe first the product of $x + y$ and $x - y$.

$$
\begin{array}{ll}
\text{Multiplicand,} & x + y \\
\text{Multiplier,} & x - y \\
\hline
& x^2 + xy \\
& \quad\;\; - xy - y^2 \\
\hline
\text{Product,} & x^2 \qquad - y^2 \qquad \text{(Answer)}
\end{array}
$$

Thus, the factors of $x^2 - y^2$ are $x + y$ and $x - y$. This example leads us to the following general rules.

9. *The product of the sum and difference of two algebraic numbers is equal to the square of the first number minus the square of the second number.*
10. *The factors of the difference of two algebraic numbers are the sum of the square roots of the numbers, and the difference of the square roots of the numbers.*

Example 10

$$a - b = \left(\sqrt{a} + \sqrt{b}\right)\left(\sqrt{a} - \sqrt{b}\right)$$

Exercises 10–4   Extract the square root:

1. $I^2 + 2IR + R^2$

2. $a^6 - 25b^2$

3. $E^4 - 9I^2$

4. $4a^2 + 12ab + 9b^2$

5. $R^2 - 1$

6. $4E^2 - 4$

7. $16t^4 - 4s^2$

8. $\frac{1}{25}x^2$

9. $\frac{1}{9}a^2 - \frac{2}{15}ab + \frac{1}{25}b^2$

10. $\frac{1}{25}x^2 + \frac{4}{5}xy + 4y^2$

11. $25I^2 - 90I + 81$

12. $36Z^2 + 12RZ + R^2$

13. $I^2 - 16I^2R^4$

14. $(a - b)^2 - 36$

15. $R^2 - (E - I)^2$

16. $x^2 - (a - b)^2$

17. $\frac{1}{9}a^2 - \frac{1}{6}ab + \frac{1}{16}b^2$

18. $\frac{1}{25}I^2 - \frac{1}{15}I + \frac{1}{36}$

19. $\frac{E^2}{R^2} + \frac{2E}{9R} + \frac{1}{81}$

20. $4\frac{Z^2}{N^2} - 16\frac{Z}{N}R + 16R^2$

▶ 10.8 Prime Factors of Polynomials

We have learned how to calculate the prime factors of numbers such as 42 by successive division. Thus, the prime factors of 42 are 2, 3, and 7. Similarly, polynomials have prime factors. Let us consider the polynomial $x^2 + cx + dx + cd$. Since x is common to the first two terms, and d is common to the third and fourth terms, we can factor x and d as follows:

$$x^2 + cx + dx + cd = x(x + c) + d(x + c)$$

Since $x + c$ is common to both terms in the right-hand member of the equation, we can factor $x + c$ as follows:

$$x(x + c) + d(x + c) = (x + c)(x + d)$$

The student may multiply $x + c$ by $x + d$ to verify that $(x + c)(x + d) = x^2 + cx + dx + cd$. By successive factoring of the polynomial, we calculated its two binomial factors. As would be anticipated, the order in which the factoring is done is inconsequential.

Example 11

$$x^2 + cx + dx + cd = x(x + d) + c(x + d)$$

and

$$x(x + d) + c(x + d) = (x + d)(x + c)$$

The factors $x + c$ and $x + d$ are called prime factors because $(x + c)(x + d)$ cannot be factored.

▶ 10.9 Factors of Trinomials

It follows from previous discussion that a trinomial can be factored if we can find two numbers whose sum is the coefficient of the second term, and whose product is the third term. Let us consider the factors of $x^2 + 5x + 6$. We wish to find two numbers whose sum is 5, and whose product is 6. Evidently, these numbers are 3 and 2. We observe that

$$(x + 3)(x + 2) = x^2 + 5x + 6$$

This example leads us to the following general rule.

11. *To factor a trinomial of the form $x^2 + bx + c$, inspect the values of the b and c to determine a pair of numbers m and n whose sum equals b and whose product equals c; in turn, the factors of the trinomial are $x + m$ and $x + n$.*

Otherwise stated,

$$x^2 + (m + n)x + mn = (x + m)(x + n)$$

To verify that the product of two binomials is equal to the trinomial, perform the multiplication of the two binomials.

Example 12

$$(3x + 1)(x + 2)$$
$$3x^2 + 6x + x + 2 = 3x^2 + 7x + 2$$

At this point, only the trial-and-error method for factoring of the foregoing type of polynomial is open to us. However, a systematic algebraic solution will be explained subsequently under the topic of *simultaneous equations*.

Exercises 10–5 Find the prime factors:

1. $x^2 - 2x - 3$

2. $x^2 + 8xy + y^2$

3. $25R^2 - 90RE + 81E$

4. $16 - 24e + 9e^2$

5. $IR - IE - ZR + ZE$

6. $C^2 + 14CL + 49L^2$

7. $1 + 4x + 4x^2$

8. $I^2 + 9IR + 18R^2$

9. $a^2 - 11ab + 28b^2$

10. $I^2 + 12IE + 36E^2$

11. $-12x + 9 + 4x^2$

12. $9R^2 - 21R + 10$

13. $4a^{2n} - 20a^n + 25$

14. $4(a - b) + (a - b)^2 + 4$

15. $4W + 28WV + 45V^2$

16. $2a^4b + a^3 + 2a^2b + a$

17. $21 + 10E + E^2$

18. $30 + b^2 - 11b$

19. $4I^2 - 12IE + 9E^2$

20. $72 + R^2 + 17R$

21. $4WV + W^2 - 77V^2$

22. $42i^5 - 28i^4e + 21i^3e^2$

23. $a(x + y) - b(x + y)$

24. $8 + 9Y^2 - 18Y$

25. $(R + E)^2 + R + E - 30$

26. $I(E - 3) - 3(E - 3)$

27. $(a - b)^2 - 2a + 2b$

28. $(a + b)^2 + 1 - 2(a + b)$

29. $2Z^2 + 3 + 5Z$

30. $e^2 + eR - 2ei - 2eR$

31. $106 - 6bR - 5a + 3aR$

32. $-13m^3 + 42m^2 + m^4$

33. $3N + 5N^2 - 14$

34. $18E^3 - 9E^2 - 9E$

35. $3a^3 - 39a^3b + 108a^3b^2$

36. $ae^2 + (a^2 + a)e + a^2$

37. $(x - y)(x + y) - 2yi - i^2$

38. $a - ab + 1 - b^2$

39. $(3E - 6)^2 - 9I^2$

40. $12x^5 - 6x^4 - 6x^3$

41. $I^2 + 6R - R^2 - 9$

42. $(a - 3)^2 - (a^2 + a + 1)^2$

43. $52ax - 8a + 28ax^2$

44. $2i^2 - 4ie + 8fe - 8f^2$

45. $E(E - 1) - R(R + 1)$

46. $aE^4 + 25R^4 - 39E^2R^2$

47. $ae + ai - be - a - bi + b$

48. $y - x - 2 + (x + y)^2$

49. $(a^2 - a)e - a^2 + ae^2$

50. $12 - 8(i^2 - i) + (i^2 - i)^2$

Summary

1. Factoring an algebraic expression entails the calculation of two or more expressions whose product is equal to the original algebraic expression.
2. Every abstract number has two square roots which are equal in absolute and opposite in sign.
3. The positive root of a number is the principal root.
4. To find the product of the sum and difference of two algebraic numbers, square the first number and subtract the square of the second number.
5. The factors of the difference of two algebraic numbers are the sum of the square roots of the numbers, and the difference of the square roots of the numbers.

Chapter 11

Operations on Algebraic Fractions

 11.1   Degrees of Monomials and Polynomials

The degree of a monomial is equal to the sum of the exponents of its literal factors. For example, I^2R is a monomial of the third degree, since the sum of the exponent 2 and the implied exponent 1 is 3. If we write IIR, the degree remains 3, and is the sum of the three implied exponents. Again, if we write $4I^2R$, the degree of the monomial remains 3 because the coefficient 4 is a numerical and not a literal factor. Observe that the degree of a monomial is simply a matter of definition, and is useful in stating rules of algebraic operations.

The monomial $2I^3R^2t$ is of the sixth degree insofar as the entire term is concerned. Note that the numerical coefficient 2 has no degree; the literal factor I is of the third degree; the literal factor R is of the second degree; the literal factor t is of the first degree.

In the case of a polynomial, the degree of the polynomial is the same as that of the monomial term of highest degree. For example, $EI = WI^2R$ is a polynomial of the third degree. Again, $2x^3y + 3x^2y^2 - 5xy + 7$ is a polynomial of the fourth degree.

Exercises 11–1   Specify the degree of each of the following equations.

1. ab^2c^3	**2.** $2x^2y^3 + 3xy$	**3.** $I^2R^2T^4 + T^{-2}$
4. $x^3y^2 + 2xy^3 + 3y^4$	**5.** $I^2R^2 + I^3R^3$	**6.** $2axy^2z + x^2yz^2$
7. $5ab + 3a^2b^2 + a^3b^3$	**8.** $e^2i + ei^2 + e^2i^2$	**9.** $2Z^2 + XR^2$
10. $\Phi^2\theta^3 + \Phi\theta^5$	**11.** $a^3y + x^2y^3 + 5xy$	**12.** $2ax^3 + a^2x^2 + ax^3$

 **11.2   Highest Common Factors of Monomials
             and Polynomials**

A *common factor* is a numerical value, literal number, or expression that can be divided without a remainder into each of two or more expressions. If there is only one common factor, it is called the *highest common factor*. In case there is more than one common factor, the highest common factor is the product of all the common factors. Let us consider the circuit depicted in Figure 11-1. The terminal resistance of this circuit can be written.

Example 1

$$\frac{2(R_3 + 2)(R_1 + R_2 + 1)}{2(R_1 + R_2 + 1) + R_3 + 1}$$

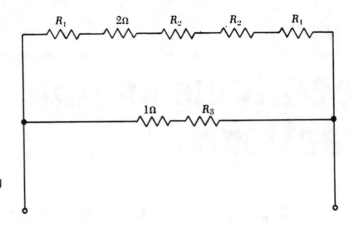

Figure 11–1
A resistive circuit entailing literal and numerical values.

It is evident by inspection that the highest common factor is the numerical value 2 in the numerator. This is the only common factor in the numerator. The denominator has no common factor, except 1, which is disregarded. Thus, we state that the highest common factor is 2 in the foregoing example.

Example 2   Note that Example 1 may be written in the form

$$\frac{2R_1R_3 + 2R_2R_3 + 2R_3 + 4R_1 + 4R_2 + 4}{2R_1 + 2R_2 + R_3 + 3}$$

The distinction between Example 2 and Example 1 is that the form of Example 1 expresses the *prime factors* which are implied in the form of Example 2. The general rules for calculating the highest common factor of two or more expressions.

1.   *Place the expressions in forms that express their prime factors.*
2.   *Inspect the prime factors to determine their common factor or factors, and choose the common factors that have the lowest exponents.*
3.   *Calculate the product of the common factors to obtain the highest common factor (HCF).*

Example 3   Find the HCF of $5x^2y^3(a - b)$ and $8xy^2(a + b)(a - b)$.

$$5 \cdot x \cdot x \cdot y \cdot y \cdot y \cdot (a - b)$$
$$2 \cdot 2 \cdot 2 \cdot x \cdot y \cdot y \cdot (a - b)(a + b)$$
$$\text{HCF} = x^2y^3(a - b)$$

Example 4   Find the HCF of $6I^2 + 12I + 6$, $15I^2 + 60I + 45$, and $3I^2 + 18I + 15$.

$$6I^2 + 12I + 6 = 2 \cdot 3(I + 1)(I + 1)$$
$$15I^2 + 60I + 45 = 3 \cdot 5(I + 1)(I + 3)$$
$$3I^2 + 18I + 15 = 3(I + 1)(I + 5)$$
$$\therefore \text{HCF} = 3(I + 1)$$

Exercises 11–2   Find the HCF of each problem.

1.   8, 32, 64
2. I^3, I^5, I^6

3. a^3b^2, a^4b^3, a^5bc^2

4. $I^2(R - E), I^3(R - E)^2$

5. $(a^2 - b^2), (a^2 + 2ab + b^2)$

6. $(I^4 - R^4), (I^4R^2 - I^2R^4)(I^6 - R^6)$

7. $2IR - 4I, IR^2 - 2IR, I^2R^2 - 4I^2$

8. $(Z + 1)(Z^2 - 1), (Z + 1)^2(Z - 1)^2, (Z^2 - 1)(Z - 1)^2$

9. $4e^4 + 32e, 3e^2 + 3e - 6, 2e^2 + 6e + 4$

10. $(2x^2 - x - 1), 4x^2 - 1, 4x^2 - 2x - 2$

11. $I^2 - r^2, I^2 - 3r + 2r^2, (I - r)^2$

12. $P^3 - 7P + 6, P^4 - 3P^3 + 6P - 4$

13. $12I^2 + 3I - 4^2, 12I^3 + 30I^2 + 12I, 32I^2 - 40I - 28$

11.3   Lowest Common Multiples of Monomials and Polynomials

A multiple of a number is another number that is divisible by the first number without a remainder. For example, I^2R is a multiple of I, because it is divisible by I without a remainder. Note that I^2R is also a multiple of I^2, and it is also a multiple of R. It is also true that I^2R is a multiple of I^2R. However, I^2R is the *smallest* literal number that is divisible by both I^2 and R without a remainder. We call I^2R the *lowest common multiple* of I^2 and R.

Example 5   Calculate the lowest common multiple (LCM) of $2I^2Rt$, II_1t^2, and $4IR$.

$$2I^2Rt = 2 \cdot I^2 \cdot R \cdot t$$
$$II_1t^2 = I \cdot I_1 \cdot t^2$$
$$4IR = 2^2 \cdot I \cdot R$$

The LCM must contain each of the three monomials, and it must contain the factors of highest degree. Accordingly, the LCM comprises $2^2, I^2, I_1, R$, and t^2, and the LCM is equal to the product of these factors of highest degree,

$$\therefore \text{LCM} = 2^2 \cdot I^2 \cdot I_1 \cdot R \cdot t^2 = 4I^2I_1Rt^2$$

These considerations lead us to the general rule for calculating the lowest common multiple of two or more numbers or expressions.

4.   *Place the numbers or expressions in a form that expresses their prime factors.*
5.   *Calculate the product of all of the different prime factors.*

Example 6   Find the LCM of $12ax^2 + 3ax - 42a, 24x^3 + 60x^2 + 24x$, and $16x^2 - 20x - 14$.

$$12ax^2 + 3ax - 42a = 3a(x + 2)(4x - 7)$$
$$24x^3 + 60x^2 + 24x = 2 \cdot 2 \cdot 3x(x + 2)(2x + 1)$$
$$16x^2 - 20x - 14 = 2(2x + 1)(4x - 7)$$
$$\therefore \text{LCM} = 6ax(2x + 1)(x + 2)(4x - 7)$$

Exercises 11–3   Find the LCM of each equation.

1. $8, 32, 64$

2. $2a^2, 3ab, 12a^2b^2$

3. $8ab^2c, 4a^2b^2c^2, 6abc^2$

4. $x^3 + x^2, x^2 + x, x^3 - x^2 + x$

5. I^2R, aI^2R^2, I^3RE

6. $NZ^2R, aeN^2Z^3, a^3N^2Z^3R$

7. $a - b, a^2 - b^2, a + 2ab + b^2$

8. $2I^2 + 4IR + 2R^2, I - R, 4I + 4R$

9. $I^2 - 4I + 3, I^2 + I - 2, 4I^2 - 4I - 24$

10. $(a^2 - 1), (a - 1)^2, (a + 1)^2, (a^2 + 1)$

11. $(P^2 - 5P + 6), (P + 3), (P^2 - 4), (P + 2)$

12. $2(i + e), 4(i^2 + 2ie + e^2), 3(i^2 + e^2)$

13. $v^3 + 2ev^2 + e^2v, 12ev^3 - 12e^3v, 2v^4 - 4ev^3 + 2e^2v^2$

14. $8a^2 - 16ab + 8b^2, (a - b)^2, 4a^2 + 8ab + b^2$

11.4   Numerators and Denominators of Algebraic Fractions

We know that a fraction is an indicated division. The term E/R is an indicated division of E by R, and $(R_1 - R_2)/(R_3 + R_4)$ is an indicated division of $R_1 - R_2$ by $R_3 + R_4$. We also know that the numerator and denominator of a fraction can be multiplied by the same number (except zero), or divided by the same number (except zero). When we multiply the numerator and denominator of a fraction by the same number, we change the *form* of the fraction, but its *value* remains the same. If two fractions have the same value, but have different forms, they are called *equivalent fractions*.

A fraction is in *lowest terms* if its numerator and its denominator have no common factor other than 1. For example, 6/9 and 2/3 are equivalent fractions, but only the fraction 2/3 is in lowest terms. Note that 6 and 9 have the common factor 3, which may be canceled in the numerator and denominator. The fraction $(R_1 + R_2R_1)/R_1$ is equivalent to the fraction $(1 + R_2)/1$, or $(1 + R_2)$, but only the latter is in lowest terms. These examples lead us to the general rules for reduction of a fraction to lowest terms.

6. *To reduce a fraction to lowest terms, place the fraction in a form that expresses the prime factors of the numerator and denominator.*

7. *Cancel the factors that are common to the numerator and to the denominator.*

Example 7

$$\frac{2ra^2 - 2rb^2}{ra^2 + 2rab + rb^2}$$

Prime factors: $\dfrac{2\,\cancel{(a-b)}\cancel{(a+b)}}{\cancel{(a+b)}\cancel{(a+b)}}$ Cancelling

$$\frac{2(a-b)}{a+b} \quad \text{(Answer)}$$

Only *common factors* may be canceled in the numerator and denominator of a fraction. Thus, we *cannot* cancel R in the numerator and denominator of the fraction $(R+R_1)/R$. Of course, we may write $(R+R_1)/R = -(R+R_1)/(-R)$, because we have used -1 as a common factor, and we multiplied both the numerator and the denominator by -1. In turn, we can cancel -1 from the numerator and denominator of $-(R+R_1)/(-R)$.

Example 8 Reduce to lowest terms.

$$-\frac{IR^2 + 2I^2R - I^3}{-I(R^2 - I^2)}$$

Prime factors: $\dfrac{-I\cancel{(R-I)}^2}{-I\cancel{(R-I)}(R+I)}$ Cancelling

$$\frac{R-I}{R+I} \quad \text{(Answer)}$$

Exercises 11–4 Reduce each of the following equations to their lowest terms.

1. $\dfrac{52x^2y^2}{74x^2y^3}$

2. $\dfrac{21a^2b^3}{63a^3b^2}$

3. $\dfrac{R_1R_2 + R_1R_3}{R_1(E+1)}$

4. $\dfrac{(v-e)^2}{v^2 - e^2}$

5. $\dfrac{ax + ay - x - y}{ax - ay - x - y}$

6. $\dfrac{-a^2 + (I-R)^2}{-(I+a)^2 + I^2}$

7. $\dfrac{e^2 - i^2}{e^2 - 2ei + i^2}$

8. $\dfrac{24x - 6x^3}{-6x^2 - 6x + 12}$

9. $\dfrac{R + 2IR - 3I^3R}{3R^2I - 4IR + R}$

10. $\dfrac{\theta^2 - \theta - 2}{\theta^3 + (\theta+1)^2 + 1}$

11. $\dfrac{x^2 + 2xy + y^2}{2(x+y)}$

12. $\dfrac{2m^2 + 4m + 2}{4m^2 + 12m + 8}$

13. $\dfrac{R^2 - 4}{R^2 - 8R + 16}$

14. $\dfrac{6a^2 - 18ab}{3b^2 - 10ab + 3a^2}$

15. $\dfrac{(a-b)^3}{a^2 - 2ab + b^2}$

16. $\dfrac{I_1^2R_2 + I_2^2R_2}{I_1R_2 + I_2R_2}$

17. $\dfrac{-2a^2 + ab - b^2}{(b-a)^2}$

18. $\dfrac{16x^2 - 32xy + 16x^2}{18x^3 - 32x}$

11.5 Conversion of Mixed Expressions to Fractions and Vice-Versa

We know how to change a mixed number into an improper fraction, and vice versa. The algebraic operation of changing a *mixed expression* into an algebraic fraction and an algebraic fraction into a mixed expression is quite similar.

Example 9 Change the following mixed expression into a fraction.

$$\beta + \frac{1}{\beta} = \frac{\beta^2 + 1}{\beta}$$

Example 10 Change the following mixed expression into a fraction.

$$\frac{1 - \partial}{\partial} + \frac{\partial^2 + \partial}{2} + \frac{\partial}{\partial^2 - 1} = \frac{(1 - \partial)(2)(\partial^2 - 1) + (\partial^2 + \partial)(\partial)(\partial^2 - 1) + \partial(\partial)(2)}{(\partial)(2)(\partial^2 - 1)}$$

$$= \frac{2\partial^2 - 2\partial^3 - 2 + 2\partial + \partial^5 + \partial^4 - \partial^3 - \partial^2 + 2\partial^2}{2\partial^3 - 2\partial}$$

$$= \frac{\partial^5 + \partial^4 - 3\partial^3 + 3\partial^2 + 2\partial - 2}{2\partial^3 - 2\partial} \quad \text{(Answer)}$$

Exercises 11–5 Change the following to a single fraction.

1. $a + \dfrac{1}{a^2}$

2. $\dfrac{i}{f} - L$

3. $\dfrac{a}{1 - a} - 3$

4. $\dfrac{t}{v} - s$

5. $\dfrac{\pi h}{b + 3} - c$

6. $\dfrac{-a}{a - 1} + 3$

7. $\dfrac{B}{B + 1} - 3$

8. $\dfrac{S + LV}{V} - (L + S)$

9. $\dfrac{R_1 + 2}{W} + R_2 + 1$

10. $\dfrac{W}{L + t} + E$

11. $\dfrac{E_1}{R_1} + E_1$

12. $a + \dfrac{x + y}{a - 1}$

13. $\dfrac{1}{a - 1} + 3 + b - \dfrac{a}{a + 1}$

14. $\dfrac{I_1 R_1}{E} + I_2 R_1 + \dfrac{I_3}{P}$

15. $\dfrac{2r^2}{1 + r} - h + r$

16. $2\pi FC + \dfrac{1}{XL} + F$

17. $\dfrac{\theta}{\Phi - 2} - \theta + \dfrac{\omega}{\lambda}$

18. $\dfrac{\tau}{2 - \pi} - \dfrac{\lambda}{2 + \beta} - \alpha$

19. $\dfrac{\Omega + \rho}{\alpha} + \dfrac{\sigma}{\eta + \lambda} + \rho$

20. $\dfrac{\alpha}{\alpha - 1} + 3\beta + \dfrac{\beta}{\beta + 1}$

Reduce to mixed expressions.

21. $\dfrac{a^3 + 2}{a^4}$

22. $\dfrac{eL - L}{e}$

23. $\dfrac{a + 2}{1 - a}$

24. $\dfrac{st - v}{v}$

25. $\dfrac{R_1 E_1 + R_1}{E_1}$

26. $\dfrac{2a + a^2 + a^3}{x - 1}$

27. $\dfrac{x^2 + 2x + 1}{(x^2 - 1)}$

28. $\dfrac{\beta^2 + \beta\alpha + \alpha^2}{\beta + 1}$

29. $\dfrac{\alpha^2 + \alpha\lambda + \lambda^2}{1 - \alpha}$

30. $\dfrac{x^2 - 4xy - 4y^2}{4xy}$

31. $\dfrac{\theta^2 - 1}{(\theta - 1)(\theta + 1)}$

32. $\dfrac{\lambda^2 - 2\lambda + 5\omega}{\lambda - 1}$

33. $\dfrac{2\tau - \tau^3 - \tau^5}{2\tau - a}$

▶ 11.6 Lowest Common Denominator of Algebraic Fractions

We are familiar with the calculation of the lowest common denominator for numerical fractions. Calculation of the LCD for two or more algebraic fractions is a similar operation. Let us consider a practical example of a circuit for which the total power dissipation is formulated, as follows.

$$W_T = \frac{V_1^2}{R_1 + 2} + \frac{V_2^2}{R_2 + 3}$$

Since the lowest common multiple of the denominators in the formula above is $(R_1 + 2)(R_2 + 3)$, this expression is the lowest common denominator. In turn, we may write the formula in the form

$$W_T = \frac{V_1^2(R_2 + 3) + V_2^2(R_1 + 2)}{(R_1 + 2)(R_2 + 3)} \tag{11.1}$$

The foregoing considerations lead us to the general rules for reduction of algebraic fractions to a fraction with the lowest common denominator.

8. *Write each algebraic fraction in a form that expresses the prime factors of the denominator.*

9. *Calculate the lowest multiple of the denominators; this expression is the lowest common denominator.*

10. *Divide the lowest common denominator by the denominator of each fraction, and multiply the numerator by this quotient.*

11. *Write each quotient as a term of the numerator (with due regard for algebraic sign), and write the lowest common denominator of the reduced fraction.*

Let us review the rules by their application to the following.

Example 11 Simplify

$$\frac{a - 2b}{a^2 - b^2} + \frac{2}{a - b}$$

Rule 8:

$$\frac{a - 2b}{(a - b)(a + b)} + \frac{2}{(a - b)}$$

Rule 9, the LCD is

$$(a - b)(a + b)$$

Rule 10:

$$\frac{a - 2b + 2(a + b)}{(a - b)(a + b)}$$

Rule 11:

$$\frac{a - 2b + 2a + 2b}{(a - b)(a + b)} = \frac{3a}{(a - b)(a + b)}$$

$$\frac{3a}{a^2 - b^2} \quad \text{(Answer)}$$

Example 12 Simplify

$$\frac{3x}{x^2 - 1} + \frac{x - 2}{x^2 + x} - \frac{4}{x + 1}$$

Prime factoring of the denominator gives

$$\frac{3x}{(x - 1)(x + 1)} + \frac{x - 2}{x(x + 1)} - \frac{4}{x + 1}$$

The LCD is then

$$x(x - 1)(x + 1)$$

then

$$\frac{(3x \cdot x) + (x - 2)(x - 1) - 4(x - 1)x}{x(x - 1)(x + 1)}$$

Expanding the numerator,

$$\frac{3x^2 + x^2 - 3x + 2 - 4x^2 + 4x}{x(x - 1)(x + 1)}$$

gives

$$\frac{x + 2}{x(x - 1)(x + 1)} \quad \text{(Answer)}$$

Exercises 11–6 Change to the lowest terms.

1. $x + \dfrac{1}{a^2}$

2. $y + \dfrac{ab}{y}$

3. $\dfrac{1}{xy} + y^2$

4. $(a + b) - \dfrac{a}{(a + b)}$

5. $\dfrac{(i - v)^2}{a} + \dfrac{i^2 + v^2}{a + 1}$

6. $\dfrac{I^2 R^2}{E} + \dfrac{(IR + 1)^2}{E + 1}$

7. $\dfrac{v_1^2}{R_1 + 3} - \dfrac{v_2^2}{R_2 - 2}$

8. $\dfrac{2a}{3} - \dfrac{3a + 1}{2} + \dfrac{a + 2}{4}$

9. $\dfrac{I - 3}{18I} - \dfrac{4I - 1}{24I} - \dfrac{1}{9}$

10. $\dfrac{e + 1}{3e^2 - 3e + 3} - \dfrac{2e}{3} + \dfrac{3e + 2}{2}$

11. $\dfrac{6R - 5}{R - 2} - \dfrac{3}{2} + \dfrac{R + 12}{4 - 2R}$

12. $\dfrac{4}{C + 1} - \dfrac{C - 2}{C^2 + C} - \dfrac{3C}{C^2 - 1}$

13. $\dfrac{1}{a + y} - \dfrac{a - x}{a(a - x)} - \dfrac{2x - a}{a^2 - x^2}$

14. $\dfrac{2E - 3R}{12RE} + \dfrac{3R - I}{6RI} - \dfrac{I - 2E}{9IE}$

15. $\dfrac{1}{6Z - 6} + \dfrac{1}{2Z + 2} - \dfrac{2Z + 1}{3Z^2 + 3Z + 3}$

16. $\dfrac{W + 6}{6} - \dfrac{W + 5}{5}$

17. $a + \dfrac{1}{a} + 2$

18. $\dfrac{2R}{I^2 - IR} + \dfrac{R^2}{I^3 - I^2R} + \dfrac{1}{I - R}$

19. $\dfrac{4}{3x - y} + \dfrac{2x + y}{y^2 - 9x^2}$

20. $\dfrac{2L - 8}{L^2 - 10L + 24} + \dfrac{L - 5}{L - 6}$

21. $\dfrac{2a}{(x - 1)} - \dfrac{4a}{x^2 + 1} + \dfrac{a}{x^2 + x}$

22. $\dfrac{1}{E_1 + E_2} - \dfrac{1}{E_1 - E_2}$

23. $\dfrac{5}{24} - \dfrac{2a^2 + 1}{3a^2} - \dfrac{a - 4}{12a}$

24. $\dfrac{E + R}{E - R} - \dfrac{E - R}{E + R}$

▶ 11.7 Addition, Subtraction, Multiplication, and Division

Operations of addition, subtraction, multiplication, and division of algebraic fractions are similar to corresponding operations on numerical fractions. For example, let us consider the addition of the two power values implicit in a series electrical circuit.

Example 13

$$W_1 = \dfrac{V_1^2}{R_1 + 2}$$

$$W_2 = \dfrac{V_2^2}{R_2 + 3}$$

In turn, the total power value is equal to the sum of P_1 and P_2.

Example 14

$$W_T = W_1 + W_2 = \dfrac{V_1^2(R^2 + 3) + V_2^2(R_1 + 2)}{(R_1 + 2)(R_2 + 3)}$$

Observe that Example 14 states the sum of the powers in Example 13. Thus, we follow Rules 8 through 11 to add algebraic fractions.

Next, let us consider the subtraction of algebraic fractions. If we are to formulate the difference between the power values in Example 14, we will evidently denote one power value as positive, and denote the other power value as negative. Then we will add these positive and negative algebraic fractions. Thus, we follow Rules 8 through 11 to subtract algebraic fractions.

In this operation we employ a minus sign merely to denote that the algebraic fraction is a *subtrahend*. It is quite possible that the numerical answer will be *negative*. This negative answer, if it occurs, *does not* imply that the resistive portion of the circuit is a power source. If the answer is negative, this sign occurs simply because the larger fractions could not be identified in the algebraic operation. Therefore, when the answer happens to be negative, we merely change the sign of the answer.

Next, consider the multiplication of algebraic fractions. This operation is similar to the multiplication of numerical fractions. We multiply the numerators together to calculate the numerator of the product; we multiply the denominators together to calculate the denominator of the product. For example, let us multiply P_1 and P_2, using the algebraic fraction noted in Example 14.

Example 15

$$W_1 W_2 = \frac{V_1^2}{R_1 + 2} \cdot \frac{V_2^2}{R_2 + 3} = \frac{V_1^2 V_2^2}{(R_1 + 2)(R_2 + 3)}$$

Next, consider the division of algebraic fractions. This operation is similar to the division of numerical fractions. One algebraic fraction will be regarded as the dividend, and the other algebraic fraction will be regarded as the divisor. In turn, we simply invert the divisor and multiply as discussed previously.

When we write an indicated division of two algebraic fractions, the expression is called a *complex fraction*. The following expression is a complex algebraic fraction.

Example 16

$$W_T = -IIR_T = \frac{E}{-R} \cdot \frac{E}{R} \left(\frac{E}{-I} + \frac{E}{I} \right) = \frac{E^2}{-R^2} \cdot 0 = 0$$

Since an algebraic term can be regarded as an algebraic fraction that has a denominator of 1, the following fractions are also called complex algebraic fractions.

Example 17

$$\frac{\dfrac{V_1^2}{R_1 + 2}}{\dfrac{V_2^2}{R_2 + 3}}$$

Invert and multiply,

$$\frac{V_1^2}{R_1 + 2} \cdot \frac{R_2 + 3}{V_2^2}$$

to obtain

$$\frac{V_1^2 (R_2 + 3)}{V_2^2 (R_1 + 2)} \quad \text{(Answer)}$$

Example 18

$$\frac{V^2}{\dfrac{V_1^2}{R_1} + 1} = \frac{V_2}{\dfrac{V_1^2 + R_1}{R_1}} = \frac{V^2 R_1}{V_1^2 + R_1} \quad \text{(Answer)}$$

Example 19

$$\frac{\dfrac{V_1}{R_1}}{V_2 + 3} = \frac{\dfrac{V_1}{R_1}}{\dfrac{V_2 + 3}{1}} = \frac{V_1}{R_1} \cdot \frac{1}{V^2 + 3} = \frac{V_1}{R_1(V_2 + 3)} \quad \text{(Answer)}$$

Division of algebraic fractions is often facilitated, as in multiplication of algebraic fractions, by inspection of numerators and denominators for common factors that can be canceled.

Exercises 11–7 Perform the indicated operations.

1. $x \cdot \dfrac{1}{a^2}$

2. $\dfrac{x}{a} \cdot \dfrac{a^2}{x^2}$

3. $\dfrac{3I^2 R}{5E} \cdot \dfrac{10RE}{I}$

4. $\dfrac{a-b}{ab} \cdot \dfrac{ab^2}{a-b}$

5. $\dfrac{x-1}{x} \cdot \dfrac{x}{x+1}$

6. $\dfrac{a^2 + 2a + 1}{x} \div \dfrac{a-1}{x^2}$

7. $\dfrac{i^2 - v^2}{i^2 v^2} \div \left(\dfrac{1}{i} + \dfrac{1}{v} \right)$

8. $\dfrac{x+2}{30} \div \dfrac{x^2 - 4}{20x}$

9. $\dfrac{a-1}{a + ab} b \cdot \dfrac{a^3 + 1}{a^3 - 1}$

10. $\dfrac{4x^3 + 4y^3}{x - 2y} \cdot \dfrac{x^2 - 4xy + 4y^2}{4x^3 - 4y^3}$

11. $\dfrac{R^2 - 14R - 15}{R^2 - 4R - 45} \cdot \dfrac{R^2 - 6R - 27}{R^2 - 12R - 15}$

12. $\dfrac{i}{(i+r)^2 - e^2} \cdot \dfrac{(i+e)^2 - r}{i^2 + ie - ir} \div \dfrac{ie - e^2 - er}{(i-e)^2 - r^2}$

13. $\dfrac{6a^2 b}{x^2 - 4y^2} \cdot \dfrac{x + 2y}{3ab}$

14. $\dfrac{2x + 7}{3} \div \dfrac{7}{21 + 2x}$

15. $\dfrac{12x^2 y^2}{7R} \cdot \dfrac{7xyz}{3R}$

16. $\dfrac{R^2 + RI + I^2}{2R^2 - RI - 3R^2} \cdot \dfrac{R^2 - I^2}{R^3 - I^3}$

17. $\dfrac{x - 4}{6x + 36} \div \dfrac{16 - x^2}{2x + 12}$

18. $\dfrac{\theta^2 - 4\theta\Phi}{\theta - \Phi} \div \dfrac{16\theta^2 \Phi^2 - \Phi^4}{4\theta^2 - 3\theta\Phi - \Phi^2}$

19. $\dfrac{4f^2 - 10fe - 6e^2}{6e^2 + 7ef - 3f^2} \cdot \dfrac{9e - 3f}{4f^2 - 2ef - 6e^2} \div \dfrac{6e^2 + ef - f^2}{4e^2 + 8ef + 3f^2}$

20. $\dfrac{2F^2 - 2F - 4}{3F^2 - 14F + 8} \left(\dfrac{12F^2 + 10F - 12}{F^2 + F - 6} \div \dfrac{4F^2 + 10F + 6}{F^2 - F - 12} \right)$

Simplify the following complex fractions.

21. $\dfrac{\dfrac{a-b}{a+b}}{\dfrac{a^2 - b^2}{a + 2ab + b^2}}$

22. $\dfrac{\dfrac{1}{i} - \dfrac{1}{e}}{\dfrac{1}{i} + \dfrac{1}{e}}$

23. $\dfrac{\dfrac{4a}{x} + \dfrac{4b}{x}}{\dfrac{2x^2}{a + b}}$

24. $\dfrac{\dfrac{R^2 + IR - 3I^2}{RI}}{\dfrac{2R - 6}{3R}}$

25. $\dfrac{\dfrac{\beta}{\beta+1}}{\dfrac{\beta}{\beta-1}}$

26. $\dfrac{\dfrac{T_j - T_a}{R_T}}{\dfrac{I_T}{T_j + T_a}}$

27. $\dfrac{\dfrac{IR + ER}{E - 1}}{\dfrac{ER}{R + E}}$

28. $E - \dfrac{R}{E - \dfrac{E}{R + E}}$

29. $\dfrac{1 + a^3}{1 - \dfrac{a}{1 + \dfrac{a}{1 - a}}}$

30. $\dfrac{\dfrac{\alpha}{1 - \alpha}}{1 - \dfrac{1 - \alpha}{\alpha}}$

31. $\dfrac{\beta - 2}{3(1 - \beta) - \dfrac{1 - \beta^3}{\beta + \frac{1}{\beta+1}}}$

32. $\dfrac{\frac{1}{2}\left(\dfrac{\lambda}{-2}\right)}{\frac{1}{2} - \dfrac{1}{1 - \frac{\lambda-2}{\lambda+2}}}$

11.8 Ratios and Quotients

The *ratio* of one number to another number is the quotient obtained by dividing the first number by the second number. Thus, a ratio is basically a fraction, such as 3/2 or x/y. It is read, "the ratio of 3 to 2," or "the ratio of x to y." It is commonly written in the form 3:2, or $x{:}y$. Concrete numbers have a ratio, provided that the concrete numbers entail the same physical units. Thus, $(3V)/(2V)$ is a ratio, and $(NV)/(MV)$ is a ratio; and the ratios are 3/2 and N/M, respectively. These are ratios simply because the physical units cancel out, leaving the quotient of two abstract numbers.

On the other hand, an indicated division of two concrete numbers that do not entail the same physical units is *not* a ratio, but a *quotient*. In spite of the fact that the indicated division of a pair of concrete numbers may be a quotient, and not a ratio, quotients are often operated on in the same basic manner as ratios. The essential distinction is that operations on ratios always lead to valid conclusions, while operations on quotients may lead either to valid or to nonphysical conclusions, depending upon the physical situation that is undergoing analysis.

11.9 Properties of Proportions

If we express two equal ratios or quotients as an equation or as a formula, we call the equation or formula a *proportion*. For example, if x/y and w/z are equal ratios, then $x/y = w/z$ may also be written $x{:}y = w{:}z$. We read this proportion "x is to y as w is to z."

Similarly, if we express two equal quotients as a formula, we call the formula a *proportion*. For example, if E_1/I_1 and E_2/I_2 are equal quotients, then $E_1/I_1 = E_2/I_2$ may also be written $E_1{:}I_1 = E_2{:}I_2$.

A practical example of voltage calculation by resistance proportions is shown in a series circuit. Suppose a 10 Ω and a 20 Ω resistor are connected in series. The voltage of each resistor can be solved by application of Ohm's law to calculate the current flow, and then by calculation of the IR drop across each resistor. However, the circuit can also be solved without calculation of the current value, by consideration of resistance proportions. Consider the fact that the current is the same in all components in a series circuit. Then:

$$I = \frac{E}{R_t} = \frac{V_1}{I} = \frac{V_2}{I}$$

$$V_1 = \frac{R_1}{R_1 + R_2}$$

$$V_2 = \frac{R_2}{R_1 + R_2}$$

Then

$$V_1 = V_2$$

$$\frac{30}{20 + 30} = \frac{20}{20 + 30}$$

$$\frac{30}{50} = \frac{20}{50}$$

The ratio of voltage across R_1 and R_2 is 3/5 to 2/5. A supply voltage of 50 volts would result in a drop of 20 v across R_2 and a drop of 30 volts across R_1.

The *principle of variation* is the same as the *principle of proportion*. A problem in variation is solved as a problem in proportion. We will use the symbol α in the following discussion, and this symbol means "varies as." We have noted the term *variable* previously. It is evident that a variable number *varies directly* as another variable number, when the ratio of any two values of the first variable number equals the ratio of the corresponding values of the other variable number. For example, if a current flows for 8 hours per day, then the total number of hours that the current flows in 5 days will be the total hours for any other number of days as the 5 is to the other number of days.

Example 20

$$\frac{40}{h} = \frac{5}{10}$$

or

$$h = 80 \text{ hours for 10 days}$$

On the other hand, a variable number *varies inversely* as another variable number when it varies directly as the *reciprocal* of the other variable number. For example, if the rate of current flow is doubled, the time required to transfer the given quantity of electricity is halved. Thus, a current flow of 1 amp transfers 10 coulombs of electrical quantity when the current has flowed for 10 sec. In turn, a current of 2 amps transfers 10 coulombs in 5 sec.

Again, a variable number can vary directly as one number and inversely as another number at the same time. For example, Coulomb's law for electrical charges states that

$$f = \frac{Q_1 Q_2}{kd^2}$$

It follows that the force between two equal charges in a vacuum varies directly as the square of the charges, and inversely as the square of the distance between them. In turn, we can write $f \alpha Q^2$ and $f \alpha 1/d^2$. Instead of using the variation symbol, we can write $f = kQ^2$, where k is a constant equal to $1/d^2$ when d is assigned a constant value. The equation $f = kQ^2$ means that f varies directly as Q^2. Again, we can write $f = c/d^2$, where c is a constant equal to Q^2 when Q is assigned a constant value. The $f = c/d^2$ means that f varies inversely as d^2.

Problems 11-1

1. Two resistors with values of 47 kΩ and 150 kΩ are connected in series, and 18 V are dropped across the 47 kilohm resistor. What is the voltage across the 150 kilohm resistor?
2. A voltage of 220 V is dropped across a 2.2 kilohm resistor and a 3.3 kilohm resistor connected in series. What voltage is dropped across each resistor?
3. A voltage of 450 V is supplied to two resistors connected in series. If 300 V drops across the 10 kilohm resistor, what is the value of the other resistor?
4. Two resistors connected in series dissipate a total power of 1500 W. If the 300 ohm resistor dissipates 900 W, what is the value of the other resistor?
5. In the formula $E_1/I_1 = E_2/I_2$, if $E_1 = 100$ v, $I_1 = 25$ mA, and $E_2 = 750$ V, what is the value of I_2?

Solve the following.

6. (a) $\dfrac{I}{6} = \dfrac{4}{3}$ (b) $\dfrac{I}{4} = \dfrac{4}{3.2}$ (c) $\dfrac{35}{4} = \dfrac{6}{E}$

    (d) $\dfrac{57}{23} = \dfrac{E}{3.5}$ (e) $\dfrac{87}{R} = \dfrac{316}{12}$ (f) $\dfrac{198}{875} = \dfrac{33}{I}$

    (g) $\dfrac{47K}{3.3K} = \dfrac{R}{220}$ (h) $\dfrac{875}{x} = \dfrac{300}{1000}$

7. One horsepower is equivalent to 746 w. What is the equivalent wattage to (a) 52 hp, (b) 0.92 hp, and (c) 2.86 hp?
8. A certain current through a 33 kilohm resistor produces 280 v; what voltage will the same current produce across a 150 kilohm resistor?
9. An area of 400 sq cm is approximately equal to 64 sq in. How many square inches are there in an area of (a) 1,120 sq cm, (b) 220 sq cm, and (c) 2,550 sq cm?

Summary

1. A common factor is a number or expression that can be divided into every term in an expression.
2. The highest common factor is the product of all common factors.
3. A multiple of a number is another number that is divisible by the first number without a remainder.
4. To reduce a fraction to the lowest terms, cancel the factors that appear in both the numerator and denominator.
5. Division of algebraic fractions is expedited by inspection of the numerator and denominator for common terms that can be canceled.
6. A ratio is basically a fraction.

Chapter 12

Fractional Equations

 12.1 Introduction

Algebraic operations on one member of a fractional equation have been discussed previously. We are now in a good position to consider the rules and postulates of operation that involve both members and the solution of a fractional equation.

Rules

1. *Clear the equation of fractions by multiplying both members by the LCD. This will allow the denominator to be canceled.*
2. *Clear all brackets and parentheses.*
3. *By transposition, collect all terms containing the unknown to one side of the equation.*
4. *Collect like terms and express the collection of terms containing the unknown in a factored form.*
5. *Solve for the unknown by dividing each member of the equation by the coefficient of the unknown.*

Example 1   Solve for x in the equation by applications of the rules.

$$5y + \frac{y}{x} = \frac{(2y+1)}{x} + 3$$

where y is 3.

Rule 1:

$$\frac{5xy + y}{x} = \frac{(2y+1) + 3x}{x}$$

Cancel the denominator by multiplying each side of the equation by x.

Rule 2:

$$5xy + y = 2y + 1 + 3x$$

Rule 3:

$$5xy - 3x = 2y + 1 - y$$

Rule 4:

$$x(5y - 3) = y + 1$$

Rule 5:

$$x = \frac{y+1}{5y-3}$$

By substituting the unknown,

$$x = \frac{3+1}{5 \cdot 3 - 3} = \frac{4}{15-3} = \frac{4}{12} = \frac{1}{3} \quad \text{(Answer)}$$

At this point we will review five basic postulates.

- A. *Equals can be added to both members of an equation.*
- B. *Equals can be subtracted from both members of an equation.*
- C. *Both members of an equation can be multiplied by equals (except zero).*
- D. *Both members of an equation can be divided by equals (except zero).*
- E. *If fractions are equal, so are their reciprocals.*

Example 2   Solve for I by the rules and postulates for equations.

$$I - \frac{25}{6} = \frac{3I}{2} - \frac{16}{3}$$

Rule 1:

$$\frac{6I - 25}{6} = \frac{9I - 32}{6}$$

Postulate C:

$$\frac{(6I - 25)\cancel{6}}{\cancel{6}} = \frac{(9I - 32)\cancel{6}}{\cancel{6}}$$

to clear the denominator.

Rule 3 by Postulate A:

$$6I - 25 + 25 = 9I - 32 + 25$$
$$6I = 9I - 7$$

Again,

$$6I - 9I = 9I - 7 - 9I$$
$$-3I = -7$$

By Postulate D,

$$\frac{-3I}{-3} = \frac{-7}{-3}$$

Finally,

$$I = \frac{7}{3} \quad \text{(Answer)}$$

Example 3   Solve the following equation for R_1.

$$\frac{1}{R_T} = -\left(\frac{R_1 + 3}{R_1}\right) + 5$$

Note: All terms in a sign of grouping preceded by a minus sign are multiplied by -1.

$$\frac{1}{R_T} = \frac{-R_1 - 3 + 5R_1}{R_1}$$

Cross multiplying,

$$R_1 = (-R_1 - 3 + 5R_1)R_T$$

$$R_1 = 4R_1R_T - 3R_T$$

$$R_1 - 4R_1R_T = -3R_T$$

$$R_1(1 - 4R_T) = -3R_T$$

$$R_1 = \frac{-3R_t}{1 - 4R_T} = \frac{3R_T}{4R_T - 1}$$

Our first step is to write the formula *clear of fractions*. This is accomplished by multiplying each term in the formula by the lowest common denominator.

Example 4

$$\frac{V_1}{I} + 3 = \frac{V_2}{I}$$

$$\frac{V_1 + 3I}{I} = \frac{V_2}{I}$$

$$V_1 + 3I = V_2$$

$$V_1 - V_2 + 3I = 0$$

Next, let us take another example that illustrates the operation of *cross multiplication*.

Example 5

$$\frac{1}{2} \cdot \frac{V_B}{I} = \frac{V_A}{I} - \frac{R_1}{2}$$

The LCD is $2I$, and we write in turn

$$\frac{V_B}{2I} = \frac{2V_A - IR_1}{2I}$$

Since we wish to calculate V_B, let us cancel 2 from both denominators and write

$$\frac{V_B}{I} = \frac{2V_a - IR_1}{I}$$

If we now *cross multiply*, we will obtain V_B directly. Cross multiplication is diagrammed as follows:

$$\frac{V_B}{I} \diagdown \frac{(2V_A - IR_1)}{I}$$

In other words, we multiply both sides by I and cancel.

$$V_B = 2V_A - IR_1$$

Note carefully that we are not permitted to cross multiply in the first formula of Example 5 because the right-hand member comprises two terms, each of which has a different denominator. However, it is permissible to cross multiply in the second formula.

The first formula in Example 5 has the common fraction $1/2$ as the coefficient of the left-hand member, and also as the coefficient of R_1. We can express the common fraction $1/2$ as the decimal fraction 0.5.

Exercises 12–1 Solve the following equations and formulas for the unknown.

1. $\dfrac{6E}{2} - 4 = 8$

2. $\dfrac{2I}{3} - 3 = \dfrac{I}{3} + 5$

3. $\dfrac{R}{3} + \dfrac{R}{2} = 4$

4. $\dfrac{3i}{4} + \dfrac{2I}{3} = 5$

5. $\dfrac{Z+1}{3} - Z = 2 - Z$

6. $\dfrac{3}{e+2} + 1 = 3$

7. $\dfrac{5}{\beta + 1} = \dfrac{6}{\beta}$

8. $\dfrac{-(1+\alpha)}{\alpha} + 2 = 3$

9. $\dfrac{1}{2c} = 3$

10. $\dfrac{3a}{2} - \dfrac{3+a}{3} = \dfrac{-6a+2}{4}$

11. $\dfrac{m+n}{m+n} + 2 = \dfrac{3}{m+n}$

12. $\dfrac{Ix+2}{3} + 2x = \dfrac{12-x}{2}$

13. $\dfrac{3(2\pi + 3e)}{2} + \dfrac{2v - 5}{3} = \dfrac{1}{2}(a+3) + \dfrac{1}{3}$

14. $\dfrac{(8x - 18)}{2} - \dfrac{3(x+1)}{3} = \dfrac{-6(2x-3)}{2}$

15. $\dfrac{4}{3\Phi + 2} + 4 = \dfrac{3}{3 - 2\Phi} - 2$

Solve the equations and formulas for the indicated letter.

16. $x_c = \dfrac{1}{2\pi fc}$ (c)

17. $\dfrac{1}{R_T} = \dfrac{1}{R_1} + \dfrac{1}{R_2} + \dfrac{1}{R_3}$ (R_T)

18. $x_L = 2\pi fL$ (L)

19. $f_0 = \dfrac{1}{2\pi\sqrt{LC}}$ (L)

20. $R = \dfrac{1}{2\pi FC}$ (F)

21. $\Phi = 1.25N^2 I$ (N)

22. $\dfrac{N_p{}^2}{N_s{}^2} = \dfrac{Z_p}{Z_s}$ (N_p)

23. $\breve{C} = \dfrac{5}{9}(F - 32)$ (F)

24. $R + r = \dfrac{Er}{e}$ (e)

25. $A = 2\pi r^2 + 2\pi rh$ (h)

26. $rt = \dfrac{a}{p} - 1$ (p)

27. $r^2 = \dfrac{m_1 m_2}{F}$ (m_1)

28. $v = \dfrac{m+M}{am}$ (m)

29. $V = \pi h^2\left(r - \dfrac{h}{3}\right)$ (r)

30. $R = \dfrac{P}{L}(L - b + a)$ (L)

31. $S = \dfrac{gt^2}{2} + v_0 t + s_0$ (g)

32. $I_C = \beta I_B + I_{CO}$ (β)

33. $W^3 = \dfrac{R(Z-1)}{L^2}$ (L)

34. $\mu = \dfrac{P(r_p + R_p)}{E_g}$ (r_p)

35. $\dfrac{1}{x} + \dfrac{1}{nx} = \dfrac{1}{f}$ (f)

36. $\dfrac{N_p}{N_s} = \dfrac{V_p}{V_s}$ (N_s)

37. $\dfrac{D}{\mu} + \dfrac{KT}{q}$ (q)

38. $\beta = \dfrac{\alpha}{1-\alpha}$ (α)

39. $A_v = \dfrac{R_L}{r_{eb}}\alpha$ (R_L)

40. $h_{21e} = \dfrac{-h_{21b}}{1 + h_{21b}}$ (h_{21b})

41. $h_{11e} = \dfrac{h_{11b}}{1 + h_{21b}}$ (h_{11b})

42. $A_i = \dfrac{-h_{21}}{1 + h_{22}R_L}$ (h_{21})

43. $\lambda = \dfrac{300 \times 10^{-6}}{f}$ (f)

44. $\mu E_g = P(r + R_p)$ (P)

45. $Z_1 = \dfrac{R_p(\mu_{eg} - e_i)}{e_1}$ (R_p)

46. $I_s = \dfrac{E_a - I_s Z_R}{a^2 Z^1}$ (I_s)

47. $\dfrac{\beta}{\Phi} + \dfrac{\beta}{\theta} + \dfrac{\beta}{\lambda} = 1$ (β)

48. $R_L = \sqrt{\dfrac{h_{11}}{h_{22}h}}$ (h_{11})

49. $S = \dfrac{R_E + R_B}{R_E + R_B(1 - \alpha)}$ (R_B)

50. $Y_n = \dfrac{y_{bc}}{1 + r_{bb}(y_{be} + y_{bc})}$ (r_{bb})

 ## 12.2 Series-Parallel Resistive Networks

The technician is often asked to analyze various combinations of series-parallel circuits such as are shown in Figure 12-1. The usual solution to these networks involves reduction of the parallel combinations to equivalent series resistances and reduction of the series equivalents to one equivalent resistance.

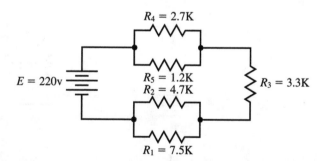

Figure 12–1
A series-parallel circuit.

A great variety of possible resistive networks prevents us from stating exact rules or procedures for their solution. We must examine each circuit and determine from our knowledge of series and parallel circuits the proper steps for solution. There are usually several different possible routes for solution, and skill in the proper selection of the simplest method will develop with practice.

Example 6   Find the total current flow in the circuit depicted in Figure 12-1. Reduce the parallel networks of R_2, R_1, and R_4, R_5 to equivalent series resistors.

$$R_T = \frac{R_1 R_2}{R_1 + R_2} + R_3 + \frac{R_4 R_5}{R_4 + R_5}$$

$$R_T = \frac{4.7\,\text{k} \times 7.5\,\text{k}}{4.7\,\text{k} + 7.5\,\text{k}} + 3.3\,\text{k} + \frac{2.7\,\text{k} \times 1.2\,\text{k}}{2.7\,\text{k} + 1.2\,\text{k}}$$

$$R_T = 2.9\,\text{k} + 3.3\,\text{k} + 0.83\,\text{k}$$

$$R_T = 7.03\,\text{k}$$

$$I_T = \frac{E_T}{R_T} = \frac{220}{7.03\,\text{k}} = 31.3\,\text{mA}$$

The solution of some series-parallel circuits may require redrawing the circuit. Sometimes labeling of junction points facilitates the process.

Example 7   Redraw the circuit in Figure 12-2, and determine I_T. Rearrangement of the circuit gives

$$R_1 = 2\,\text{k}\Omega \parallel 12\,\text{k}\Omega \parallel 6\,\text{k}\Omega + 5\,\text{k}\Omega^*$$

which gives

$$R_T = \frac{1}{G_1 + G_2 + G_3} + 5\,\text{k}\Omega$$

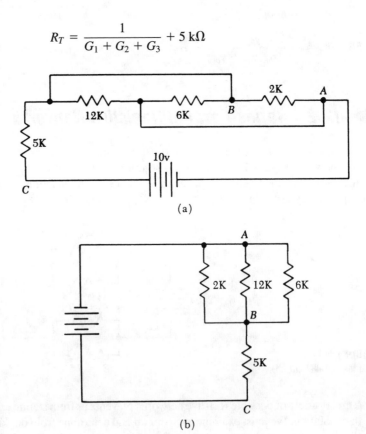

(a)

(b)

Figure 12–2
Series-parallel connected
circuits.

*$\parallel$ indicates in parallel to.

or

$$R_T = \cfrac{1}{\cfrac{1}{R_1} + \cfrac{1}{R_2} + \cfrac{1}{R_3}} + 5 \text{ k}\Omega$$

$$R_T = 1.33 \text{ k}\Omega + 5 \text{ k}\Omega$$

(Answer)

$$R_T = 1.33 \text{ k}\Omega + 5 \text{ k}\Omega = 6.33 \text{ k}\Omega \quad \text{(Answer)}$$

The solution of series-parallel networks very often will require application of one or more circuit laws, such as Ohm's law, Kirchhoff's law, or the power laws to find unknown values.

Example 8 Find the total current and voltage values in the circuit depicted in Figure 12-3. Note that 40 V are present across R_4; therefore, there are 40 V dropped across R_5 and 20 V dropped across each R_2 and R_3.

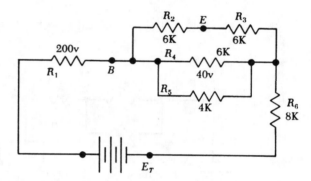

Figure 12–3
A series-parallel connected circuit.

$$I_5 = \frac{40 \text{ V}}{4 \text{ k}\Omega} = 10 \text{ mA}$$

$$I_4 = \frac{40 \text{ V}}{6 \text{ k}\Omega} = 3.3 \text{ mA}$$

$$\vdots$$

$$I_2 = I_3 = \frac{20 \text{ V}}{6 \text{ k}} = 3.3 \text{ mA}$$

then

$$I_T = 10 \text{ mA} + 6.67 \text{ mA} + 3.33 \text{ mA} = 20 \text{ mA}$$

$$V_1 = I_T \times 10 \text{ k} = 200 \text{ V}$$

$$V_6 = I_T \times 8 \text{ k} = 160 \text{ V}$$

and

$$E_T = V_1 + V_p + V_6$$

$$E_T = 200 \text{ V} + 40 \text{ V} + 160 \text{ V} = 200 \text{ V}$$

Problems 12–1

1. Solve each of the networks in Figure 12-4 for the total resistance values.

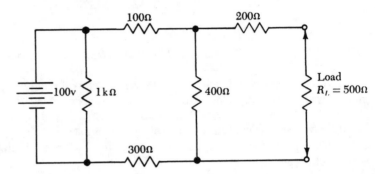

Figure 12–4
A series-parallel connected circuit.

2. From the circuit in Figure 12-5, determine the values of R_T, I_3, I_2, and I_T.

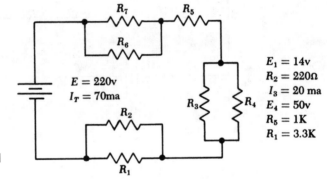

Figure 12–5
A series-parallel connected circuit.

3. From the circuit in Figure 12-6, determine the values of I_T, I_3, I_2, and R_T.

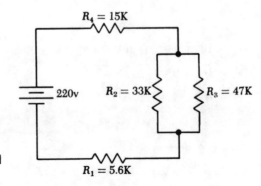

Figure 12–6
A series-parallel connected circuit.

4. From the circuit in Figure 12-7, determine the values of R_T, I_T, I_2, I_3, I_5, and I_6.

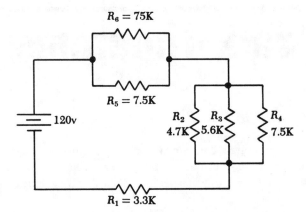

Figure 12–7
A series-parallel connected
circuit.

5. Determine the value of R_T when each resistor in the circuit of Figure 12-5 is 1 kΩ.

Summary

1. The equal sign in an equation is the balance indicating that one side is exactly equal to the other.
2. Equal values can be added or subtracted from both sides of an equation.
3. Both sides of an equation can be divided or multiplied by the equal values.
4. Both sides of an equation can be raised to the same power.

Chapter 13

Network Theorems

 13.1 Introduction

Network theorems can save a great amount of time in solution of complex electrical circuits. For example, we earlier derived the *maximum power transfer theorem*, which states that the maximum power is transferred to the load when the load is equal to the source resistance.

 13.2 Thevenin's Theorem

One of the most important circuit theorems entails a simple equivalent circuit with a voltage source and a series resistance. This theorem is called Thevenin's or Pollard's theorem. This theorem is stated as follows: *Any linear network of resistances and sources, if viewed from two points in the network, can be replaced by an equivalent resistance R_{th} in series with an equivalent voltage source V_{th}.* The process whereby a Thevenin equivalent circuit is derived for a given network is illustrated in Figure 13-1.

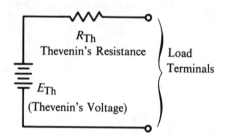

Figure 13–1
Thevenin's equivalent
circuit.

We shall assume that the circuit in Figure 13-2 is to be analyzed by Thevenin's theorem. Our problem consists of evaluating the circuit for V_{th} and R_{th}. We proceed as follows:

1. *Disconnect the section of the circuit considered to be the load—R_L in Figure 13-2(a).*
2. *By measurement or calculation, determine the voltage that would appear between the load terminals with the load disconnected (open-circuit voltage), for terminals X and Y. This open-circuit voltage is called the Thevenin's voltage, V_{th}.*
3. *Replace each voltage source within the circuit by its internal resistance. A constant-voltage source is replaced by a short circuit as in Figure 13-2(b).*
4. *By measurement or calculation, determine the resistance that would be "seen" looking back into the open load terminals; this is the Thevenin's resistance, R_{th} in Figure 13-2(b).*
5. *Draw the equivalent circuit consisting of R_L and R_{th} in series, connected across V_{th}, as shown in Figure 13-2(c). Use the voltage divider rule to solve for the load voltage and current.*

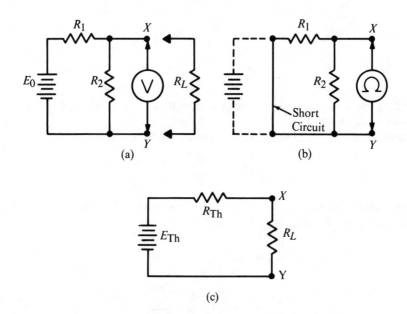

Figure 13–2 Deriving a Thevenin's equivalent circuit.

Example 1 Determine the value of the load voltage V_L and load current for the circuit shown in Figure 13-3 using Thevenin's theorem.

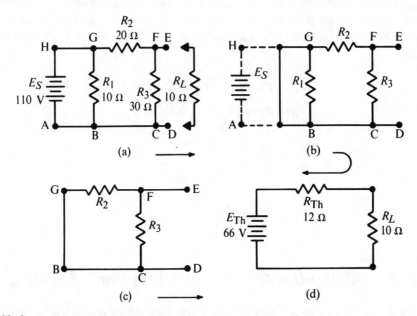

Figure 13–3 Evaluation of the equivalent circuit.

1. Disconnect the load as shown in Figure 13-3(a).
2. Compute the open-circuit voltage, V_{th}, between terminal E and D with R_L disconnected.

$$V_L = \frac{R_3}{R_2 + R_3} E_S$$

$$V_L = \frac{30}{20 + 30} 110 = 66 \text{ V} \qquad \textbf{(13.1)}$$

3. Replace the voltage source with a short circuit, since it is constant-voltage source, as shown Figure 13-3(b).
4. Determine R_{th}, which consists of R_2 and R_3 in parallel (R_1 is shorted):

$$R_{th} = \frac{R_2 R_3}{R_2 + R_3} = \frac{20 \times 30}{20 + 30} = \frac{600}{50} = 12 \ \Omega \qquad \textbf{(13.2)}$$

5. Draw the equivalent circuit and connect the load resistor as shown in Figure 13-3(d).
6. Calculate the load voltage by the voltage-divider rule:

$$V_L = \frac{R_L}{R_{th} + R_L} V_{th}$$

$$= \frac{10}{12 + 10} 66 = \frac{660}{22} = 30 \text{ V}$$

$$I_L = \frac{V_L}{R_L} = 30 \frac{v}{10 \ \Omega} = 3 \text{ Ampere}$$

The economy of time and effort in calculation of load voltage and current is obvious. However, the answers obtained concern only the load. For example, Thevenin's solution tells us that the load voltage is 30 volts, the load current is 3 amperes, and the load power is 90 watts. On the other hand, we can conclude nothing concerning the power supplied by the source.

We can, however, apply the calculated values of load current and voltage to the original circuit to aid in the evaluation of other circuit values. With the values of V_L and I_L calculated by Thevenin's theorem, placed on the schematic in Figure 13-3, the other circuit values are obvious:

$$V_{R3} = V_L = 30 \text{ V}$$

$$V_{R2} = E_S - V_L = 110 - 30 = 80 \text{ V}$$

$$V_{R1} = E_S = 110 \text{ V}$$

$$I_3 = \frac{V_{R3}}{R_3} = \frac{30}{30} = 1 \text{ amp}$$

$$I_2 = \frac{V_{R2}}{R_2} = \frac{80}{20} = 4 \text{ amps}$$

$$I_1 = \frac{V_{R1}}{R_1} = \frac{110}{10} = 11 \text{ amps}$$

▶ *13.3 Voltage-Divider Rule and Thevenin's Theorem*

The voltage across any resistor of a series circuit can be solved by a simple ratio which we shall call the voltage-divider rule. The formula is derived for the fact that the current is the same in all parts of a series circuit:

$$I = \frac{E}{R_T} = \frac{V_1}{R_1} = \frac{V_2}{R_2}$$

The open-circuit voltage is the voltage drop across R_3 with the load removed.

$$V_{th} = \frac{R_3}{R_1 + R_3}E_S = \frac{30}{20 + 30}110 = 66 \text{ V}$$

From the Thevenin's circuit in Figure 13-3(c).

$$R_{th} = R_2 \parallel R_3 = 20 \parallel 30 = 12 \text{ }\Omega$$

$$I_T = \frac{E_{th}}{R_{th} + R_L} = \frac{66}{12 + 10} = \frac{66}{22} = 3 \text{ A}$$

$$V_L = I \times R_L = 3 \times 10 = 30 \text{ V}$$

▶ 13.4 Norton's Theorem

Thevenin's theorem states that the parameters of an equivalent circuit employ a constant-voltage source with an equivalent series resistance. On the other hand, Norton's theorem states the parameters of the equivalent circuit as a constant-current source with an internal parallel conductance. The current source (I_N) is found by measuring the current that results when the load is shorted. The internal conductance is found by removing the load, shorting all voltage sources, opening all current sources and looking back into the circuit from the load terminals.

Norton's theorem, as Thevenin's theorem is a tool for solving complex circuits or representing control devices in circuit problems. The choice of one over the other depends upon the type of circuit and the reader's personal choice.

Example 2 Apply Norton's theorem to determine the voltage across the load in Figure 13-4.

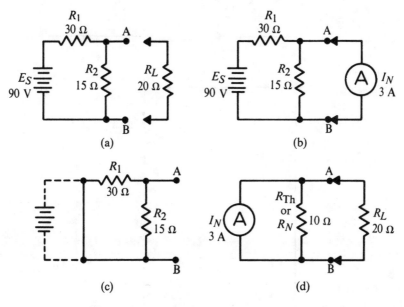

Figure 13-4 A circuit reduced to the equivalent Norton's circuit for a given load.

1. The short-circuit current is found from Figure 13-4(b).

$$I_N = \frac{E_S}{R_1} = \frac{90 \text{ V}}{30 \text{ } \Omega} = 3 \text{ A}$$

From Figure 13-4(c) the internal resistance or conductance is:

$$R_N = R_1 \parallel R_2 = 30 \text{ } \Omega \parallel 20 \text{ } \Omega = 12 \text{ } \Omega$$

$$G_N = \frac{1}{R_N} = \frac{1}{10} = 0.1 \text{ Siemen}$$

Either Norton's or Thevenin's theorem will be applicable to solve most circuits. However, it has been the author's experience that students grasp the concept of Thevenin's theorem easier than Norton's theorem and most will choose to solve a circuit by the application of Thevenin's and convert to Norton's when that solution is required.

Thevenin's or Norton's theorems allow any component or components to be assumed by the load. Therefore, the theorems may be applied to a very complex circuit more than once, as illustrated in Example 3.

Exercises 13–1   Find the load current for the following circuits using both Thevenin's and Norton's theorems.

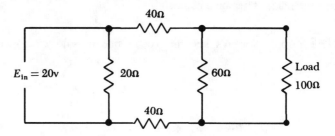

Figure 13–5
A complex circuit to be solved by Thevenin's and Norton's theorems.

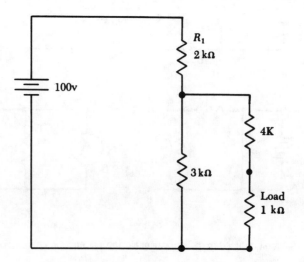

Figure 13–6
A complex circuit to be solved by Thevenin's and Norton's theorems.

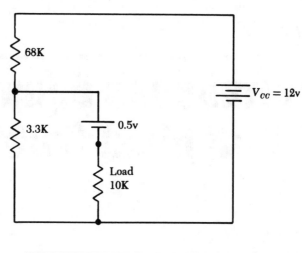

Figure 13–7
A complex circuit to be solved by Thevenin's and Norton's theorems.

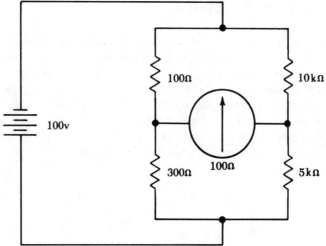

Figure 13–8
A complex circuit to be solved by Thevenin's and Norton's theorems.

Problems 13–1

1. Solve for the load current and load voltage in the circuit shown in Figure 13-5.
2. Using the circuit shown in Figure 13-6 solve for load current and load voltage.
3. Using the circuit shown in Figure 13-7 solve for equivalent Thevenin's and Norton's circuits and find I_L.
4. Using the bridge circuit in Figure 13-8 solve for the load current and voltage for the 100 Ω meter.

Summary

1. Thevenin's theorem is a mathematical model that reduces a complex circuit to an equivalent source voltage with a series resistance for a given load.
2. Norton's theorem is a mathematical model that reduces a complex circuit to an equivalent current source with a parallel conductance for a given load.
3. Any part of a complex circuit can be assumed to be the load in application of Thevenin's or Norton's theorems.

Chapter 14

Simultaneous Equations and Graphical Solutions

▶ *14.1 Introduction*

Presentation on a graph or chart is a forceful method of representing data and presenting the relationships between data. We observe graphs of events, such as the rise and fall of the stock-market prices, general rise of the cost of living, and rate of coal consumption versus fuel-oil consumption, almost daily. These relations can be effectively presented on a graph, even though they cannot be mathematically formulated. Obviously, data with only two variables that are related to each other, so that whatever value one has, the other has some corresponding value, are the best adapted for presentation on a graph. We say of this type of data that one unknown is a function of the other unknown; y is a function of x, or x is a function of y (Figure 14-1).

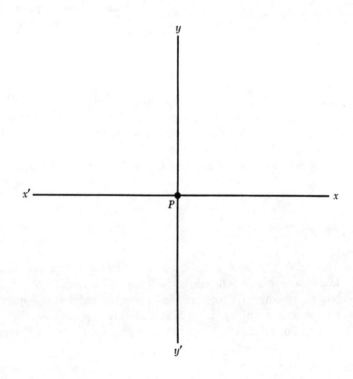

Figure 14–1
Rectangular (Cartesian)
coordinate system.

14.2 Rectangular or Cartesian Coordinates

A rectangular system of coordinates is formed by drawing two perpendicular lines through a point, P, in Figure 14-1. The horizontal line XX' is called the *abscissa* or x axis, and the vertical line YY' is called the *ordinate* or y axis.

The location of any point on the plane is known if the distance from the origin is known for both the x axis and the y axis. By convention, the abscissa is measured to the right if positive, and to the left if negative. The ordinate is measured upward if positive, and downward if negative. Location of any point on a graph is fixed by its Cartesian coordinates, as shown in Figure 14-2, with the examples of points $(-2, -3)$, $(-3, 5)$, $(4, 4)$, and $(2, -6)$.

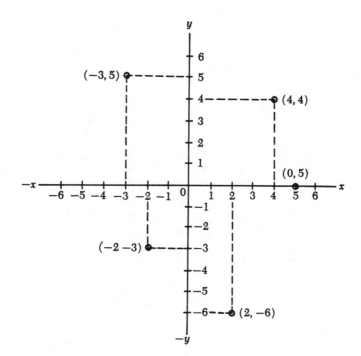

Figure 14–2
Location of four points on
a coordinate system.

14.3 Graphical Solutions of Equations

Previous discussion has touched upon *indeterminate* equations and formulas. For example, $I_1 + I_2 = 7$ is an indeterminate formula; if $I_1 = 1$, then $I_2 = 6$; if $I_1 = 2$, then $I_2 = 5$; if $I_1 = 3$, then $I_2 = 4$, and may be plotted as a straight line as shown in Figure 14-3. However, if it is stipulated that $I_1 = 2I_2$, these indeterminate formulas are called *simultaneous* formulas.

In other words, both formulas are satisfied by values for I_1 and I_2. In situations of this kind, we have learned to substitute from one formula into the other to solve the problem that is presented. Thus, to continue the foregoing example, we would substitute $2I_2$ for I_1, and write $I_2 + 2I_2 = 7$, or $I_2 = 7/3$ and $I_1 = 14/3$. Now, let us visualize what we have done.

Figure 14-4 depicts plots of $I_1 + I_2 = 7$, and $I_1 = 2I_2$. Observe that the two plots have a point in common at $I_1 = 14/3$ and $I_2 = 7/3$. In other words, these numerical values for I_1 and I_2 satisfy our pair of indeterminate formulas simultaneously. This might seem to be a trivial observation, but it has great practical importance.

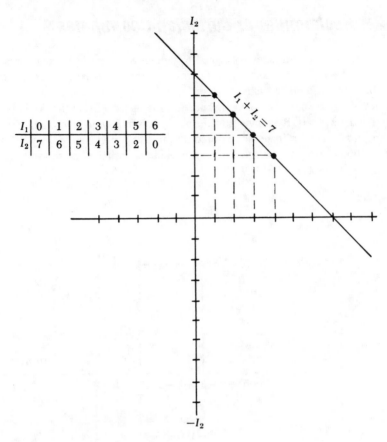

I_1	0	1	2	3	4	5	6
I_2	7	6	5	4	3	2	0

Figure 14–3 Graph of a straight line.

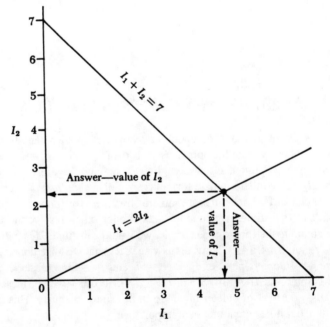

Figure 14–4
Graphical solution of a pair
of simultaneous equations.

Exercises 14–1 Graph the following functions.

1. $x + y = 6$
2. $2x + y = 5$
3. $3I_1 + 3I_2 = 10$
4. $3x + 4y = 12$
5. $x - 3y = 6$
6. $3I_2 - 4I_1 = 12$

 ## 14.4 Algebraic Solution of Simultaneous Equations

We have seen that a pair of simple simultaneous equations can often be solved easily by substitution. Another approach consists of algebraic addition and subtraction.

Example 1 Consider the pair of simultaneous equations

$$2x + 3y = 4$$
$$3x - 3y = 6$$

sum $\qquad 5x = 10$

If we add these two equations, we obtain

$$5x = 10$$
$$x = 2$$

To calculate the value of y, we can substitute the number 2 for x into either of the simultaneous equations.

Alternatively, we can subtract one of the simultaneous equations from the other.

Example 2

$$3x - 4y = 12$$
$$3x - 6y = -18$$

Difference $\qquad 2y = 30$

$$y = 15$$

Substituting the number 15 for y in the equation $3x - 4y = 12$ gives

$$3x - 4 \cdot 15 = 2$$
$$3x = 12 + 60$$
$$3x = 72$$
$$x = 24$$

Then

$$3x - 4y = 12$$
$$3 \cdot 24 - 4y = 12$$
$$72 - 4y = 12$$
$$4y = 60$$
$$y = 15$$

In both of the foregoing examples, we have added or subtracted to *eliminate either one of the variables*. Let us next consider how to eliminate one of the variables in the following pair of simultaneous equations.

$$3x + 2y = 6$$
$$-x + 3y = 1$$

Since the coefficients for x and y are different in this pair of equations, we cannot eliminate either of the variables by direct addition or subtraction. However, if we multiply through the second equation by 3, we observe that the coefficient of x will be the same in both equations. Now we can eliminate the variable x by adding the simultaneous equations.

Example 3

$$3x - 5y = 10$$
$$x - y = 2$$

Note that to permit elimination of x by addition, we multiplied $x - y = 2$ by -3 to obtain $-3x + 3y = -6$.

$$3x - 5y = 10$$
$$-3x + 3y = -6$$

Sum $\quad - 2y = 4$

$$y = -2$$

To calculate the value of x, we may substitute the number -2 for y in either of the simultaneous equations and obtain $x = 0$.

Now let us apply our new knowledge to analyze the electrical circuit depicted in Figure 14-5. The circuit has a possibility of two current loops and different directions that we may assign to

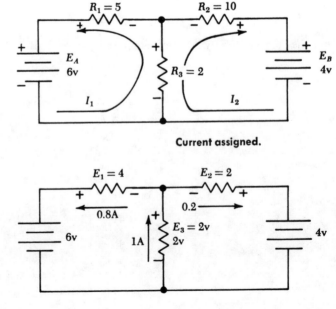

Current assigned.

Figure 14–5
Solving circuit currents by linear equations.

the two currents. Our choice of direction is of no consequence to the final answer, as an improper choice of direction will yield the proper magnitude and be indicated by a negative sign. *One note of caution:* once the direction of current is selected, we must write our formula of voltage drops with the resultant polarities.

Example 4 Find the voltage drop across R_1, R_2, and R_3, and check the answer.

Procedure: Assign the currents I_1 and I_2, and indicate the polarity of the voltage drops as in Figure 14-5(a). The left loop results in the equation

$$0 = 6 - E_{R_1} - E_{R_2} = 0 = 6 - 2I_1 - 2(I_1 + I_2)$$

Collecting terms and changing signs,

$$6 = 7I_1 + 2I_2$$

The right loop yields

$$0 = E_B - E_{R_3} - E_{R_2}$$
$$0 = 4 - (2I_1 + 2I_2) - 10I_2$$

Collecting terms and changing signs,

$$4 = 2I_1 + 12I_2$$

We now have two formulas with two unknowns and, therefore, may use any of the methods studied to solve for I_1 and I_2. Let us use the subtraction method by multiplying Formula 1 by 6 to eliminate I_2.

$$\begin{array}{lr} 6 \times \text{Formula 1,} & 36 = 42I_1 + 12I_2 \\ \text{Formula 2,} & 4 = 2I_1 + 12I_2 \\ \hline \text{Difference gives} & 32 = 40I_1 \end{array}$$

Solving for I_1 gives

$$I_1 = 0.8 \text{ A}$$

Substituting I_1 into Formula 2 gives

$$4 = 2(0.8) + 12I_2$$
$$I_2 = \frac{2.4}{12} = 0.2 \text{ A}$$

Testing our currents in Figure 14-5(b) indicates that the voltage drops $E_{B1} + E_3 = 6$ V, and $E_{B2} + E_{B3} = 4$ V, proving our calculations are correct.

Exercises 14–2

1. Determine the magnitude and direction of the current through R_L in the circuit in Figure 14-6.
2. Determine by Kirchhoff's law I_L and V_L in the circuit in Figure 14-7.
3. Determine the value of I_3 in the circuit in Figure 14-8, if $R_1 = R_4$.
4. Determine E_L and R_L for the circuit in Figure 14-9 using Kirchhoff's law.

5. Determine the voltage drop across R_2 and R_3 in the circuit in Figure 14-7 when $R_1 = R_2 = R_L$.

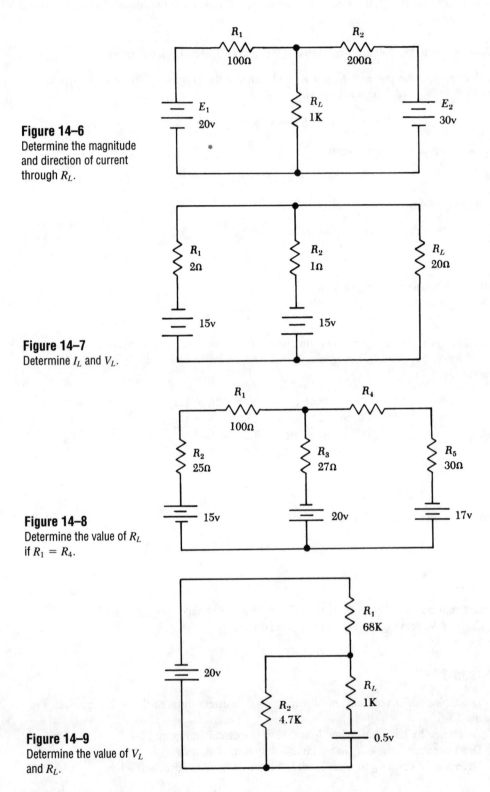

Figure 14–6
Determine the magnitude and direction of current through R_L.

Figure 14–7
Determine I_L and V_L.

Figure 14–8
Determine the value of R_L if $R_1 = R_4$.

Figure 14–9
Determine the value of V_L and R_L.

14.5 Three Simultaneous Equations in Three Unknowns

Completing solutions of a set of simultaneous equations requires that we have as many equations as there are unknowns. Thus, if we have two simultaneous equations in three unknowns, our solution will be an indeterminate equation. But if we have three simultaneous equations in three unknowns, we can obtain a determinate solution.

Three simultaneous equations may be solved in successive steps. First, we must choose any pair of equations, and eliminate one of the variables, such as z. Then, we must choose another pair of equations, and eliminate the same variable. This procedure yields a pair of simultaneous equations in two unknowns, such as x and y. In turn, we can eliminate one of these unknowns, such as y. We have then solved for the value of x. This value of x can then be substituted into one of the equations in two unknowns to solve for y. Finally, we can substitute the known values of x and y into one of the original equations to solve for z. This process is very time consuming and tedious. A more productive method will be considered here.

14.6 Basic Determinants

Determinants are arrays of numbers that can often save considerable labor in the solution of simultaneous equations. Determinants are used to solve *linear* simultaneous equations or formulas, which means that the exponents of the variables must be 1. This is the most common situation in routine engineering work. Considerable labor is saved, in particular, when there are more than three unknowns with which to contend. The determinant for a set of simultaneous formulas is written by noting the coefficient of the various unknowns.

Example 5   Set up the determinants of the equations below and solve for I_1 and I_2.

$$E_1 = a_1 I_1 + b_1 + I_2$$
$$E_2 = a_2 I_1 + b_2 I_2$$

Write the simultaneous equations with the same unknowns in the same column, as if we were to proceed by addition or subtraction. The coefficients of I_1 are a_1 and a_2; the coefficients of I_2 are b_1 and b_2, and the constants are E_1 and E_2. The determinant for the denominator of our formula is written by placing the coefficients of each variable in the denominator of the determinant. The determinant of the numerator is *the same as the denominator, except the constant replaces the variable of the term for which we are solving*. The determinant for the denominator is the same for each unknown. We now cross multiply.

Note, that the upward diagonal is subtracted.

$$I_1 = \frac{\begin{vmatrix} E_1 & b_1 \\ E_2 & b_2 \end{vmatrix}}{\begin{vmatrix} a_1 & b_1 \\ a_2 & b_2 \end{vmatrix}} = \frac{E_1 b_2 - E_2 b_1}{a_1 b_2 - a_2 b_1}$$

We write the determinant and solve for I_2.

$$\frac{\begin{vmatrix} a_1 & E_1 \\ a_2 & E_2 \end{vmatrix}}{\begin{vmatrix} a_1 & b_1 \\ a_2 & b_2 \end{vmatrix}} = \frac{a_1 E_2 - a_2 E_1}{a_1 b_2 - a_2 b_1}$$

The unknowns for the denominator need be calculated only once. Let us apply our knowledge to a practical problem.

Example 6

$$2I_1 + 3I_2 = 20$$

$$5I_1 + 7I_2 = 30$$

$$I_1 = \frac{\begin{vmatrix} 20 & 3 \\ 30 & 7 \end{vmatrix}}{\begin{vmatrix} 2 & 3 \\ 5 & 7 \end{vmatrix}} = \frac{140 - 90}{14 - 15} = \frac{50}{-1} = -50 \text{ A}$$

$$I_2 = \frac{\begin{vmatrix} 2 & 20 \\ 5 & 30 \end{vmatrix}}{\begin{vmatrix} 2 & 3 \\ 5 & 7 \end{vmatrix}} = \frac{60 - 100}{14 - 15} = \frac{-40}{-1} = 40 \text{ A}$$

Checking by substituting in the first equation

$$2(-50) + 3(40) = 20$$

$$-100 + 120 = 20$$

$$20 = 20$$

Determinants with three unknowns are very common in electronics and other sciences. However, determinants of more than three unknowns have very little application except for the solution of complex circuits.

$$a_1 I_1 + b_1 I_2 + c_1 I_3 = E_1$$
$$a_2 I_1 + b_2 I_2 + c_2 I_3 = E_2$$
$$a_3 I_1 + b_3 I_2 + c_3 I_3 = E_3$$

$$I_1 = \frac{\begin{vmatrix} E_1 & b_1 & c_1 \\ E_2 & b_2 & c_2 \\ E_3 & b_3 & c_3 \end{vmatrix}}{\begin{vmatrix} a_1 & b_1 & c_1 \\ a_2 & b_2 & c_2 \\ a_3 & b_3 & c_3 \end{vmatrix}}, I_2 = \frac{\begin{vmatrix} a_1 & E_1 & c_1 \\ a_2 & E_2 & c_2 \\ a_3 & E_3 & c_3 \end{vmatrix}}{\begin{vmatrix} a_1 & b_1 & c_1 \\ a_2 & b_2 & c_2 \\ a_3 & b_3 & c_3 \end{vmatrix}}, I_3 = \frac{\begin{vmatrix} a_1 & b_1 & E_1 \\ a_2 & b_2 & E_2 \\ a_3 & b_3 & E_3 \end{vmatrix}}{\begin{vmatrix} a_1 & b_1 & c_1 \\ a_2 & b_2 & c_2 \\ a_3 & b_3 & c_3 \end{vmatrix}}$$

Solving for I_1 we add the sum of the products of the three downward diagonals, and subtract the sum of the products of the three upward diagonals.

$$I_1 = \frac{\begin{vmatrix} E_1 & b_1 & c_1 \\ E_2 & b_2 & c_2 \\ E_3 & b_3 & c_3 \end{vmatrix}}{\begin{vmatrix} a_1 & b_1 & c_1 \\ a_2 & b_2 & c_2 \\ a_3 & b_3 & c_3 \end{vmatrix}} = \frac{(E_1 b_2 c_3 + E_2 b_3 c_1 + E_3 c_2 b_1) - (E_3 b_2 c_1 + E_2 b_1 c_3 + E_1 c_2 b_3)}{(a_1 b_2 c_3 + a_2 b_3 c_1 + a_3 c_2 b) - (a_3 b_2 a_1 + a_2 b_1 c_3 + a_1 c_2 b_3)}$$

It is left to the reader to solve the determinates for I_2 and I_3. The results of the denominator for each unknown is the same as for I_1. The numerators are:

$$I_2 = \frac{(a_1 E_2 c_3 + a_2 E_2 c_1 + a_3 c_2 E_1) - (a_3 E_2 c_1 + a_2 E_1 c_3 + a_1 c_2 E_3)}{}$$

and

$$I_3 = \frac{(a_1 b_2 E_3 + a_2 b_3 E_1 + a_3 E_2 b_1) - (a_3 b_2 E_1 + a_2 b_1 E_3 + a_1 E_2 b_3)}{}$$

Note that each term includes 3 factors, one from each equation, for example, $a_3 E_2 c_1$. Let us now apply these techniques to three simultaneous formulas.

Example 7

$$1I_1 + 2I_2 + 3I_3 = 2$$
$$3I_2 + 5I_3 = 6$$
$$2I_1 \quad\quad + 2I_3 = 4$$

We must put the coefficients of all variables into the determinant, even those that are zero. Then follow the example above and draw the diagonals, exactly.

$$I_1 = \frac{\begin{vmatrix} 2 & 2 & 3 \\ 6 & 3 & 5 \\ 4 & 0 & 2 \end{vmatrix}}{\begin{vmatrix} 1 & 2 & 3 \\ 0 & 3 & 5 \\ 2 & 0 & 2 \end{vmatrix}} = \frac{(40 + 0 + 12) - (36 + 0 + 24)}{(20 + 0 + 6) - (18 + 0 + 0)} = \frac{52 - 60}{26 - 18} = \frac{-8}{8} = -1$$

$$I_2 = \frac{\begin{vmatrix} 1 & 2 & 3 \\ 0 & 6 & 5 \\ 2 & 4 & 2 \end{vmatrix}}{8} = \frac{(20 + 0 + 12) - (36 + 20 + 0)}{8} = \frac{32 - 56}{8} = -\frac{24}{8} = -3$$

$$I_3 = \frac{\begin{vmatrix} 1 & 2 & 2 \\ 0 & 3 & 6 \\ 2 & 0 & 4 \end{vmatrix}}{8} = \frac{(24 + 0 + 12) - (12 + 0 + 0)}{8} = \frac{36 - 12}{8} = \frac{24}{8} = 3$$

Problems 14–1 Solve the following problems by the method of your choice.

1. Find three currents such that their sum is 60 amp; $1/2$ of I_1 + $1/3$ of I_2 and $1/5$ of I_3 is 19: and twice I_1 with three times the remainder when I_3 is subtracted from I_2 is 50.

2. An electronics technician has 70 semiconductor devices, diodes, junction transistors, field-effect transistors, and operational amplifiers. $1/4$ of the number are diodes, $1/6$ are junction transistors, and $1/2$ of the field-effect transistors equals 26. If he subtracts the sum of the junction transistors and diodes from the number of operational amplifiers, the remainder is 10. How many semiconductors of each kind does he have?

3. An IBM PC computer and an Apple computer can do a piece of work in .8 minutes; the PC computer and a Philco 110 computer can do the work in .10 minutes; the IIE computer

and the 110 computer can do the work in .10 minutes. In how many minutes can each of the computers do the same work alone?

4. A high-fidelity "buff" has a 120-watt amplifier, a 60 watt amplifier, and a 20 watt amplifier on a speaker. If the speaker is put with a 120 watt amplifier, the system will be worth $10.00 more than both the other amplifiers. If the speaker is put with the 60 watt amplifier, the system will be worth $20.00 more than twice the value of the 20 watt amplifier. If the speaker is connected with the 20 watt amplifier, the value of the system will be 1/5 that of the 120 watt amplifier alone and 1/3 of the 60 watt amplifier alone. What is the value of each unit if their total cost is $400.00?

5. A parts dealer bought 8 receivers, a number of phonographs, and 100 record albums for $2,500.00. The number of phonographs was equal numerically to 4 times the price of a record album, and an album and a receiver cost $5.00 less than 1/5 the cost of all the phonographs. Find the cost of a receiver, an album, and the total number of phonographs if one phonograph cost $40.00.

6. Three electronics technicians, A, B, and C, jointly purchased a quantity of transistors for $900.00. The sum paid by A increased by 2/3 of the sum B paid, and diminished by 1/2 the sum of what C paid, was $320.00. One-half of what A paid added to 1/4 of what B paid and 1/5 of what C paid was $279.00. What amount did each of the electronics technicians pay?

7. The value of a resistor is expressed by three digits whose sum is 10. The sum of the first and last digits is 2/3 of the second digit, and if 198 is subtracted from the number, the digits will be inverted. What is the value of the resistor?

8. Three students A, B, and C, purchased a number of diodes, transistors, and resistors at the same rate. A paid $4.20 for 7 diodes, 5 transistors, and 3 resistors; B paid $3.40 for 9 diodes, 4 transistors, and 2 resistors; C paid $3.25 for 5 diodes, 2 transistors, and 3 resistors. What was the price of each component?

9. While taking inventory of his parts, an electronics technician finds that he has 84 resistors, capacitors, and fuses worth $42.00. He also finds that 1/3 of his resistors and 1/4 of his capacitors are worth $6.50. How many of each component did he have if the value of the resistors is 50 cents each, the value of the capacitors is 25 cents each, and the fuses are valued at one dollar each?

Summary

1. The relationship between current and voltage in a resistive circuit can be portrayed by a linear equation.
2. Simultaneous linear equations can be solved by: substitution, subtraction or addition and determinants.
3. Complex simultaneous linear equations may be readily solved by the use of the determinants.
4. The use of determinants to solve linear equations requires careful placement of the coefficient of each variable.

Chapter 15

Complex Algebra

 15.1 Introduction

We know that various sorts of numbers are necessary to calculate electric and electronic circuit action. For example, we have made extensive application of the following sorts of numbers.

1. *Natural numbers:* 1, 2, 3, 4, 5, 6, 7, ...
2. *Prime numbers:* 1, 2, 3, 5, 7, 11, 13, ...
3. *Counting numbers:* 0, 1, 2, 3, 4, 5, 6, ...
4. *Negative numbers:* $-1, -2, -3, -4, -5, \ldots$
5. *Rational numbers:* $1.2, -1/3, 2/2, -9/3, \ldots$
6. *Irrational numbers:* $\sqrt{2}, -\sqrt{2}, \sqrt{2}.\pi, \ldots$

The *real numbers* consist of the rational numbers and the irrational numbers. In other words, all of the six sorts of numbers listed above are real numbers. However, the real numbers alone are insufficient to calculate electric and electronic circuit action in many practical situations. Therefore, we must consider another sort of numbers that are called *imaginary numbers*. The sum or difference of a real number and an imaginary number is called a *complex number*.

The most basic imaginary number is written $\sqrt{-1}$. In literal form, $\sqrt{-1}$ is written i; engineers prefer to use the form j, so that there is no possibility of confusing this symbol with the symbol for electric current. Note that the imaginary unit is defined as the *positive* square root of -1.

The square of every real number is either positive or zero. Therefore, $\sqrt{-1}$ is not less than zero, equal to zero, or greater than zero. Since $\sqrt{-1}$ is a unique sort of number, it was called an imaginary number by its discoverer. We will find, however, than an imaginary current is physically real. The value of an imaginary current can be measured with an ammeter, and an imaginary current will give us a shock just as a real current does.

15.2 Geometrical Representation of Imaginary Numbers

The geometrical representation of the real numbers and the imaginary numbers depicted in Figure 15-1 provides a visualization of the relation between these two sorts of numbers. If points along the horizontal axis of a coordinate system are represented by real numbers, then points along the vertical axis may be represented by imaginary numbers. Let us see why this is so.

Consider the geometrical interpretation of the mathematical operation in which we multiply $+1$ successively by $\sqrt{-1}$. If we multiply $+1$ on the horizontal axis in Figure 15-1 by $\sqrt{-1}$, the answer is $+\sqrt{-1}$, which is located on the vertical axis. Multiplication of the distance from 0 to $+1$ by $\sqrt{-1}$ rotates the interval through 90°, and the answer is the distance from 0 to $+\sqrt{-1}$. Next, if we multiply $+\sqrt{-1}$ on the vertical axis by $\sqrt{-1}$, the answer is -1 on the horizontal axis. Otherwise stated, multiplication of the distance from 0 to $+\sqrt{-1}$ by $\sqrt{-1}$ rotates the interval through 90 degrees, and the answer is the distance from 0 to -1.

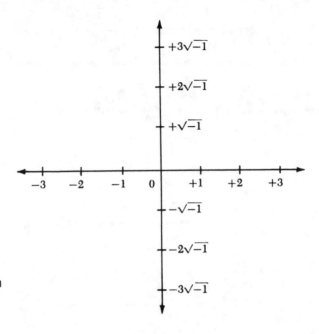

Figure 15–1
Geometrical representation
of the real numbers and
the imaginary numbers.

If we multiply -1 by $\sqrt{-1}$, the distance from 0 to -1 on the horizontal axis is rotated $90°$, and the answer is the distance from 0 to $-\sqrt{-1}$ on the vertical axis. Finally, if we multiply $-\sqrt{-1}$ by $\sqrt{-1}$, the distance from 0 to $-\sqrt{-1}$ is rotated $90°$, and the answer is the distance from 0 to $+1$ on the horizontal axis. Thus, multiplication by $\sqrt{-1}$ four times brings us back to the starting point.

The $\sqrt{-1}$ is called the j operator, because it operates to rotate a numbered interval 90 degrees counterclockwise on a coordinate system.

 15.3   Pure Imaginary Numbers

A *pure imaginary number* is symbolized $\sqrt{-n}$, where $-n$ is a negative real number. Let us consider how pure imaginary numbers are added. It is evident from Figure 15-1 that $\sqrt{-1}$ plus $2\sqrt{-1}$ is equal to $3\sqrt{-1}$. Next, let us add $\sqrt{-4}$ to $\sqrt{-25}$; this operation denotes $2\sqrt{-1} + 5\sqrt{-1} = 7\sqrt{-1}$. Again, let us add $\sqrt{-4}$ to $\sqrt{-3}$; this operation yields $2\sqrt{-1} + \sqrt{3} \cdot \sqrt{-1} = j(2 + \sqrt{3})$.

Pure imaginary numbers are subtracted in the same general manner. For example, let us subtract $\sqrt{-3}$ from $\sqrt{-2}$; this operation denotes $j\sqrt{2} - j\sqrt{3} = \sqrt{-1}(\sqrt{2} - \sqrt{3}) = j(\sqrt{2} - \sqrt{3})$.

Next, let us consider the multiplication of pure imaginary numbers. If we multiply $\sqrt{-4}$ by $\sqrt{-25}$, this operation denotes $(j2) \cdot (j5) = j^2 10 = -1 \times -10 = +10$. Note $j^2 = (\sqrt{-1})^2 = -1$.

Imaginary numbers are divided as shown in Example 1.

Example 1   Consider the division of $\sqrt{-4}$ by $\sqrt{-25}$.

$$\frac{\sqrt{-4}}{\sqrt{-25}} = \frac{j\sqrt{4}}{j\sqrt{25}} = \frac{j2}{j5} = 0.4$$

The js in the numerator and denominator cancel in the equation in Example 1.

Exercises 15–1 Draw a geometrical representation of real and imaginary numbers, and locate the following.

1. $-j$ 2. j^2 3. $\sqrt{-3}$ 4. $\sqrt{-4}$

5. $-\sqrt{-9}$ 6. $-\sqrt{25}$ 7. $-6\sqrt{-4}$ 8. $3j^2$

9. $-4j^2$ 10. $-5j^3$ 11. $3j^3$ 12. $2j^5$

13. $-4j^5$ 14. $3j^6$

▶ 15.4 Complex Numbers and Operations

A *complex number* is defined as the sum or difference of a real number and an imaginary number; thus, a complex number has the form $1 + j3$. The indicated addition of 2 and $j3$, for example is written $2 + j3$; the indicated subtraction of 2 and $j3$ is written $2 - j3$. Neither the sum nor the difference is a real number, nor is the sum or difference a pure imaginary number. Instead, this sort of indicated sum or difference is a complex number. Just as real numbers and pure imaginary numbers have geometric interpretations, so does a complex number have a geometric interpretation. Figure 15-2 depicts the geometric representation of the complex numbers $2 + j3$ and $-2 - j3$.

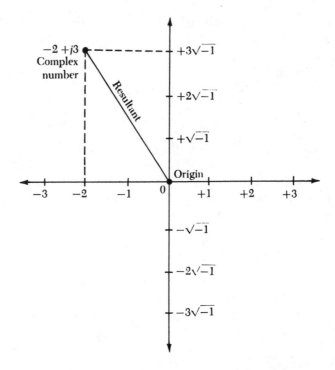

Figure 15–2
Geometric representation of the complex number $2 + j3$ and $-2 - j3$ and their resultant.

Every complex number has a resultant, which is the distance from the origin to the complex number in the complex plane, as depicted in Figure 15-2. The resultant is the absolute value of the complex number.

The absolute values of complex numbers are scalars. However, the *directed magnitude* is not a scalar. There is a sense, though, in which the resultant of a complex number has direction as well as magnitude. Thus, the resultant has a certain angle with respect to the axis of real numbers. We call this resultant and direction a *vector*.

To add a pair of complex numbers, we add the *real parts* of the complex numbers, and we add the *imaginary parts* of the complex numbers; finally, we express the sums as a complex number.

Example 2 Add $-2 + j3$ to $1 - j5$.

$$\begin{array}{r} -2 + j3 \\ \underline{1 - j5} \\ -1 - j2 \end{array}$$

To subtract one complex number from another, we proceed as in simple algebra, merely keeping the real numbers and imaginary numbers separate.

Example 3 Subtract $2 + 5\sqrt{-1}$ from $-5 + 3\sqrt{-1}$.

$$\begin{array}{r} -5 + j3 \\ \underline{-2 - j5} \\ -7 - j2 \end{array}$$

We subtract the real parts of the complex numbers, and we subtract the imaginary parts of the complex numbers; finally, we express the difference as a complex number.

To multiply one complex number by another, we proceed as in simple algebra, merely observing that j^2 is equal to $(\sqrt{-1.})^2$ or -1.

Example 4 Multiply $2 + j3$ by $3 - j2$.

$$\begin{array}{r} 2 + j3 \\ \underline{3 - j2} \\ 6 + j9 \\ -j4 - j^2 6 = 6 + 6 + j5 = 12 + j5 \end{array}$$

To divide one complex number by another, it is often convenient to *rationalize* the denominator. This is an operation that changes the denominator into a real number or into a rational number. For example, let us simplify the fraction

$$\frac{2}{3 + j5}$$

We will rationalize the denominator. We will perform this operation by multiplying both numerator and denominator by the *conjugate* of the denominator. The conjugate of a binomial is the same binomial with the middle sign changed. Thus, the conjugate of $3 + j5$ is $3 - j5$.

Example 5

$$\frac{2(3 - j5)}{(3 + j5)(3 - j5)} = \frac{6 - j10}{9 - j15 + j15 - j^2 25}$$

$$= \frac{6 - j10}{9 - (-1)25} = \frac{6}{34} = -\frac{j10}{34} = 0.176 - j0.294$$

Example 6

$$\frac{2(3 - j5)}{(3 + j5)(3 - j5)} = \frac{6 - 2j5}{9 - 3j5 + 3j5 - j^25}$$

$$= \frac{6 - 2j5}{9 - (-1)5} = \frac{6 - 2j5}{14} = \frac{6}{14} - \frac{j(2)(2.36)}{14} = 0.428 - j4.72$$

We may also simplify a fraction with an imaginary number in the denominator by changing the denominator from rectangular for such as $10 + j10$ to the polar form of $14.1/45°$. Note that the rectangular form represents the two sides of a right triangle and the polar form represents the hypotenuse and the angle of the triangle.

Example 7   Simplify the following fraction.

$$\frac{10}{1 + j1} = \frac{10}{1.414/45°} = 7.07/ - 45°$$

Exercises 15–2   Perform the indicated operations.

1. $3\sqrt{-3} + 4\sqrt{-3}$

2. $\sqrt{-25} + \sqrt{-16}$

3. $2\sqrt{-4} + 3\sqrt{-16}$

4. $-j2 + j3$

5. $3 + j^25$

6. $-5 - j^25$

7. $-3 + j^48$

8. $15 + j^33$

9. $12 - 3j^3$

10. $-3 - j^35$

11. $2\sqrt{-4} \times \sqrt{-9}$

12. $\sqrt{-50} \times \sqrt{-32}$

13. $j^2 \times j3$

14. $\sqrt{-25} \times \sqrt{-9}$

15. $-\sqrt{-5} \times \sqrt{-9}$

16. $j^2 \times j^26$

17. $j^25 \times j^46$

18. $j3 \times j^34$

19. $-j^3 \times j^3$

20. $-j^23 \times (-j^25)$

21. $2\sqrt{-9} - \sqrt{-4}$

22. $5\sqrt{-50} - 3\sqrt{-16}$

23. $12\sqrt{-3} - 2\sqrt{-9}$

24. $6\sqrt{-4} - 2\sqrt{-2}$

25. $j^23 - j^45$

26. $j^32 - j^54$

27. $j^35 - j^52$

28. $-j^26 - j^43$

29. $\dfrac{\sqrt{-4}}{\sqrt{-16}}$

30. $\dfrac{\sqrt{-9}}{\sqrt{-3}}$

31. $\dfrac{\sqrt{-125}}{\sqrt{-50}}$

32. $\dfrac{\sqrt{-50}}{\sqrt{-8}}$

33. $\dfrac{j^3}{j^2}$

34. $\dfrac{12}{j^22}$

35. $\dfrac{25}{j^3 10}$ **36.** $\dfrac{64}{-j^3}$

37. $\dfrac{j^2 15}{25}$ **38.** $\dfrac{-j^4 12}{3j}$

39. $\dfrac{-j^5 2}{-j^3 4}$ **40.** $\dfrac{-j^3 25}{-j 10}$

Problems 15–1

1. Explain the meaning of the j operator.
2. What is the conjugate of $-2 + j\sqrt{3}$?
3. Could you receive a shock from an imaginary electric current?
4. What is the conjugate of $2 + \sqrt{3}$?
5. What is the conjugate of $-2 - \sqrt{3}$?

Summary

1. Complex numbers are utilized in electronics to indicate current and voltage values that are ± 90 degrees out of phase and to solve circuit problems involving these relationships.
2. Operator $j(\sqrt{-1})$ is utilized to indicate ± 90 degree relationships. Plus $\sqrt{-1}(j)$ indicates a leading $90°$ relationship and $-\sqrt{-1}(-j)$ indicates a lagging $90°$ relationship.

Chapter 16

Logarithms

 ## 16.1 Introduction

Multiplication, division, involution, and evolution of large numbers is greatly facilitated by the use of *logarithms*.

The rules for the application of logarithms are as follows:

1. *Multiplication is accomplished by addition of logarithms.*
2. *Division is accomplished by subtraction of logarithms.*
3. *Involution (rising to a power) is accomplished by multiplying logarithms.*
4. *Evolution (extracting a root) is accomplished by dividing logarithms.*

These operations entail expression of a given number as a power of some number chosen as the base.

The common system of logarithms utilizes 10 as the base. Natural logarithms, on the other hand, utilize 2.7182818285..., which is usually rounded off to 2.71828 or 2.71, as their base. The base of common logarithms is identified as $\log_{10}$ or log. The base of natural logarithms is symbolized by e or epsilon (ϵ) and identified as 1_n. Both common and natural logarithms are important in applications of circuit action.

16.2 Common Logarithms

The common logarithm of a chosen number is the power to which 10 must be raised to equal the chosen number. Any arithmetical number is equal to some power of 10.

Example 1

$$10^0 = 1, \text{ or } \log 1 = 0$$
$$10^1 = 10, \text{ or } \log 10 = 1$$
$$10^2 = 100, \text{ or } \log 100 = 2$$
$$\log 34 \approx 1.53148$$

Note, that the log of 34 is 1.53148, and the log is approximate. It could be carried out as many decimal places as we please. Thus, we perceive that some logs are exact, such as $\log 100 = 2$, but that other logs are inexact.

Next, let us observe the notation for expressing the log of a decimal fraction.

Example 2

$$0.1 = \frac{1}{10^1} \text{ or } 10^{-1}; \ \log \text{ of } 10^{-1} = -1$$

Again,

$$0.001 = 10^{-3}; \log .001 = -3$$

Logarithms are most easily found by the use of a scientific calculator. To find a log with a calculator, for example, to find the log of 2000 with the calculator:

1. Place the number 2000 in the calculator.
2. Press "log."

The answer appears,

$$\log 2000 = 3.3010$$

Exercises 16–1 Practice finding logarithms to the base 10 with the calculator.

1. 4	**2.** 8	**3.** 40	**4.** 800
5. 0.4	**6.** 78,000	**7.** 0.0003	**8.** 10,000

 16.3 Antilogarithms

The antilogarithm of a logarithm results in the number from which the logarithm was derived. Determining the antilogarithm is the reverse operation of finding the logarithm of a number. For example, let us find the antilogarithm of 2.4969.

1. *Place the logarithm in the calculator.*
2. *Press* second *or INV then press "log."*

Example 3

$$N = 10^2 \times 10^{.4969}$$

$$\log N = 2.4969$$

$$\text{antilog } 2.4969$$

$$\text{antilog } N \approx 314$$

 16.4 Multiplication and Division by Logarithms

The logarithm of the product of two numbers is equal to the sum of the logarithms of the numbers. This follows logically when we remember that the logarithm is the power of the base. For example, to multiply powers of 10, we add the exponents.

Example 4

$$10^3 \times 10^5 = 10^{3+5} = 10^8$$

then:

$$N = 4.53 \times 5270$$

$$\log N = \log 4.53 + \log 5270$$

$$\log N = 0.6561 + 3.7218 = 4.3779$$

$$\text{antilog } N = \text{antilog } 4.3779 = 23,873 \ldots$$

To divide a number by another number using logarithms, we subtract the logarithm of the divisor from the logarithm of the dividend to obtain the logarithm of the answer.

Example 5

$$N = \frac{25,864}{782}$$

$$\log N = \log 25,864 - \log 782 = 4.412 - 2.893 = 1.519$$

$$N = \text{antilog } N \approx 33.04$$

The calculator makes multiplication and division of numbers by the use of logarithms almost obsolete.

 16.5   Evolution of Numbers by Logarithms

To extract the root of a number by means of logarithms, we must divide the logarithm of the number by the index root. This gives us the logarithm of the root of the number. In turn, the antilog yields the root.

Example 6   Find the value of $(42,875)^{1/3}$.

$$N = (42,875)^{1/3}$$

$$\log N = 1/3 \log 42,875 \approx 1/3(4.63) \approx 1.54$$

$$N = \text{antilog } 1.54 \approx 34.9$$

 16.6   Applications of Common Logarithms

The common logarithm of the ratio of two power values is a number called *bels*. One-tenth of a bel is equal to a *decibel*, or dB. Thus, the number of bels difference in level between a value of W_1 watts and W_2 watts is expressed

$$\text{bels} = \log_{10} \frac{W_2}{W_1} \qquad (16.1)$$

or the number of decibels difference in level between W_2 and W_1 is

$$\text{Decibels} = 10 \log_{10} \frac{W_2}{W_1} \qquad (16.2)$$

Example 7   Find the decibel gain of an amplifier with 100 watts of output power and 20 milliwatts of input power.

$$\text{dB} = 10 \log \frac{P_0}{P_{in}} = 10 \log \frac{100W}{0.02W} =$$

$$\text{dB} = 10(\log 100 - \log 0.02) = 10[2 - (-1.7)] = 10(3.7) \cong 37$$

$$A_p \cong 37 \text{ dB}$$

The decibel gain of an amplifier or system can be found from the ratio of input current and output current or input voltage and output voltage. The formulas are developed as follows:
We know that:

$$dB = 10 \log P_o/P_{in} \qquad (16.3)$$

then:

$$dB = 10 \log \frac{V_o^2/R_o}{V_{in}^2/R_{in}}$$

$$dB = 10 \log \frac{V_o^2/R_{in}}{V_{in}^2/R_o}$$

$$dB = 10 \log \left(\frac{V_o}{V_{in}}\right)^2 + 10 \log \frac{R_{in}}{R_o}$$

Note: $\log \left(\frac{V_o}{V_{in}}\right)^2 = 2 \log \frac{V_o}{V_{in}}$. Finally:

$$dB = 20 \log \frac{V_o}{V_{in}} + 10 \log \frac{R_{in}}{R_o} \qquad (16.4)$$

When the input and output resistance are of the same value the last term in the foregoing formulas will reduce to zero. Then:

$$dB = 20 \log \frac{V_{out}}{V_{in}} \text{ for voltages} \qquad (16.5)$$

And:

$$dB = 20 \log \frac{I_{out}}{I_{in}} \text{ for currents} \qquad (16.6)$$

Care must be taken when calculating dB gains of systems in which the input and output resistances are not equal as extreme error in calculations can occur.

Example 8 Find the decibel gain of an amplifier having equal input and output resistances, when an input of 30 mV produces an output of 24 V.

$$dB \ A_v \text{ gain} = 20 \log \frac{24 \text{ V}}{0.03 \text{ V}} = 20(\log 24 - \log 0.03)$$

$$dB = 20[1.380 - (-1.523) \approx 20(2.903) \approx 58.06]$$

The voltage gain is approximately 58.06 dB.

▶ 16.7 Decibel Reference Levels

Decibels refer to power *ratios*. In turn, decibels can be utilized as a measure of absolute magnitude if a *reference level* has been established. The reference level is arbitrary.

A common reference level for decibel measurements is 0.600 W into 600 Ω. In turn, the statement "20 dB (0 dB = 6 mW)" is a measure of power level in milliwatts. In other words, 0 dB is defined to correspond to 6 mw; in turn, 20 dB corresponds to a power level of 100×6 mW = 600 mW.

As proof observe:

$$dB = 10 \log \frac{P_o}{P_{in}}$$

$$dB = 10 \log \frac{600 \text{ mW}}{6 \text{ mW}}$$

$$dB = 10 \times \log 100$$

$$dB = 10 \times 2 = 200$$

In telephone and telecommunication engineering, a reference level of 6 mW 500 Ω is generally used, although you will sometimes find a reference level of 6 mw into 600 Ω.

A *dBm measurement* is a decibel measurement of a sine-wave voltage made with a decibel meter having a reference level of 1 mW into 600 Ω.

A *VU (volume unit)* is a decibel of a non-sinusoidal voltage. A sinusoidal wave represents a simple-single harmonic motion. On the other hand, a non-sinusoidal wave is an irregular variation, such as a chord produced by an organ. A VU measurement is always made with a reference to 1 mW into 600 Ω.

VU meters are found to be of maximum utility in monitoring audio-frequency signals corresponding to speech and music.

Decibel measurements made with a reference to power levels are straightforward. On the other hand, decibel measurements made with reference to current or voltage *must* be made with reference to the *same* load resistance, such as 500 Ω, 600 Ω, or some other fixed value.

Exercises 16–2 Find the decibel gain of the following circuits assuming that $R_o = R_{in}$.

1. $P_o = 2 \text{ W}, P_{in} = 20 \text{ mW}$
2. $V_o = 10 \text{ V}, V_{in} = 10 \text{ } \mu\text{V}$
3. $I_o = 1 \text{ A}, I_{in} = 100 \text{ nA}$
4. $P_o = 40 \text{ W}, P_{in} = 10 \text{ mW}$
5. $V_o = 60 \text{ mV}, V_{in} = 100 \text{ } \mu\text{V}$
6. $I_o = 1 \text{ A}, I_{in} = 1 \text{ } \mu\text{A}$

Find the unknown in the following.

7. $P_o = 22 \text{ W}$ across 600 Ω, what is the dBm measurement?
8. $P_o = 60 \text{ W}$ across 1 kΩ, what is the dBm measurement?
9. $V_{in} = 100 \text{ } \mu\text{V}$ across 1 kΩ, $V_o = 2 \text{ V}$ across 40 Ω. What is the dB gain of the amplifier?
10. An emitter follower circuit has an input resistance of 1 MΩ and an output resistance of 100 Ω; what is the dB gain of the circuit if the input voltage = the output voltage = 1 volt?

 ## 16.8   Frequency Response and Decibel Gain

Any circuit containing reactance and resistance will not have the same response at all frequencies. An *RL* or *RC* circuit can be used to *discriminate* against either high or low frequencies.

Measurement of frequency discrimination requires a suitable reference value called *cutoff frequency* (f_{co}). The term cutoff frequency is misleading, because the current in a series *RL* circuit, for example, does not stop abruptly at some upper frequency limit. Instead, the current value decreases gradually. In turn, the cutoff frequency is defined as the frequency at which:

1. *True power in the circuit has one-half of its maximum possible value.*
2. *The cutoff frequency is −3 dB.*
3. $X_L = R \text{ or } X_C = R$.

4. $V_L = V_R$ or $V_C = V_R = 0.707E$.
5. *The phase angle of the circuit is $\pm45°$.*

The formula for cutoff frequency in an *RL* circuit is derived as follows.
Since $X_L = R$ at cutoff frequency:

$$X_L = R$$

and

$$2\pi fL = R$$

then

$$f_{co} = \frac{R}{2\pi L} \tag{16.7}$$

In an *RL*-series circuit, the resistance is not affected by frequency variation. However, the value of X_L is a direct function of frequency. At zero frequency (DC), an ideal inductor has no opposition to current flow, and the current is limited only by the circuit resistance.

Example 9   Determine the circuit current when a 1000-Ω resistor is connected in series with a 15.59-mH ideal inductor and a 100-V AC source at 0 Hz, 500 Hz, and 1 kHz.

$$f_c = \frac{R}{2\pi L} = \frac{1000}{6.28 \times 0.0159} \approx \frac{1000}{0.1} \approx 10 \text{ kHz}$$

Since the maximum possible power is developed at zero frequency, the true power has decreased to one-half at 1 kHz. If we were taking the output across the resistor the output voltage would be at 0.707 of the maximum output. The voltage across the capacitor would also be 0.707 of maximum because $\theta = -45°$.

In a series RC circuit the reactance of the capacitor decreases with an increase in frequency while the value of the resistor is not affected. An output across the capacitor would result in a decrease in voltage with frequency; while an output across the resistor would result in an increase in voltage. We can develop the formula for cutoff frequency since $X_C = R$ at f_{co}.

Example 10   Consider an RC circuit that is used as a filter.

$$X_C = \frac{1}{2\pi fC} = R$$

Then

$$f_{co} = \frac{1}{2\pi RC} \tag{16.8}$$

▶ 16.9   Filter Concepts and the Bode Plot

The reactance of a capacitor varies inversely to frequency and therefore, can be used with a resistor in a series to produce a circuit in which the output voltage is dependent upon the frequency of the input voltage. An example of this action is shown in Figure 16-1. An output across the capacitor decreases as frequency increases. Conversely, the output across the resistor increases as frequency increases.

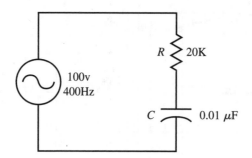

Figure 16-1
(a) An RC filter circuit, and
(b) The resulting frequency plots.

A plot of the low-pass filter is shown in Figure 16-2(a). The cutoff frequency is shown at −3 dB. If we make a further plot of the filter action we would observe that the *roll-off* would be at a rate of −6 dB per octave or −20 dB per decade. Note, that a *decade is times ten and an octave is times two*.

A *Bode plot* is made by estimating the filter action. We mark the cutoff frequency at −0 dB on the vertical, locate the −20 dB point on the graph at 10 and connecting the roll-off line. The estimated plot is accurate except between the 0 dB and the −3 dB points, an area that is of little interest to the technician or engineer.

Most high-pass and low-pass filters are constructed using capacitors and resistors rather than inductors and resistors because of the size and cost of the latter.

Example 11 Calculate the cutoff frequency of the circuit in Figure 16-1.

$$f_{co} = \frac{1}{2}\pi RC$$

$$f_{co} = \frac{0.159}{RC} = \frac{0.159}{2 \times 10^4 \times 1 \times 10^{-8}}$$

$$f = 795 \text{ Hz}$$

Problems 16-1 Find the cutoff frequency of the following and make a Bode plot of each.

1. A series RC circuit with the output taken across the capacitor when $R = 10 \text{ k}\Omega$ and $C = 100 \text{ nF}$.
2. A series RC circuit with the output taken across the resistor when $R = 1 \text{ k}\Omega$ and $C = 2 \mu F$.
3. A series RC circuit with the output taken across the capacitor when $R = 1 \text{ k}\Omega$ and $C = 159 \text{ pF}$.
4. Determine the value of the capacitance in a series RC circuit that will result in a cutoff frequency of 10 kHz when the value of the resistance is 15.9 kΩ.
5. What is the value of the resistance that will result in a cutoff frequency of 100 kHz when the capacitance is 100 pF. Plot the Bode plot for the circuit.

▶ 16.10 Band-Pass Filter Circuits

Transistor amplifiers (Figure 16-3) have series and parallel capacitances that result in the circuit being both a low-pass and a high-pass filter. The frequency response of the circuit is comprised of a low-pass filter and a high-pass that results in a band pass filter.

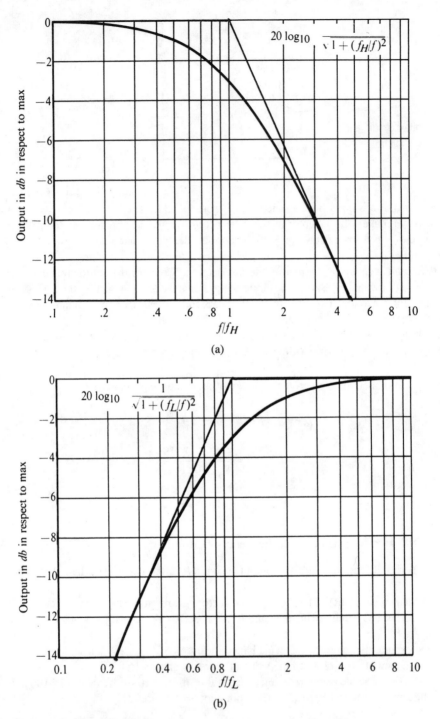

Figure 16–2 A Bode plot of the frequency versus the output power in dB with the output taken across R in the circuit in (a).

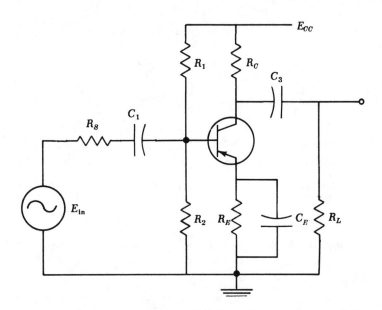

Figure 16–3
A common-emitter amplifier.

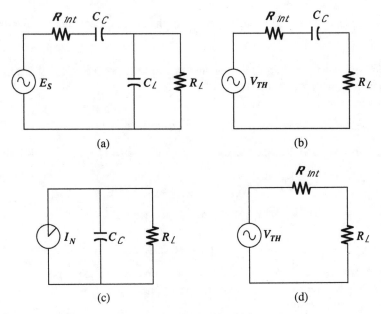

(a)   (b)

(c)   (d)

Figure 16–4 An equivalent frequency response circuit, (c) mid-frequency circuit (d) low-frequency circuit, (e) high-frequency circuit.

Capacitor C_1, C_2, and C_E are responsible for the low-frequency responses of the amplifier. For low-frequency analysis these capacitors are consolidated into a single capacitor as depicted in Figure 16-4(c). For high-frequencies analysis the transistor and wiring capacitances are consolidated into a parallel capacitance as shown in Figure 16-4(d). A Norton's equivalent current circuit is used for the analysis.

At mid-frequencies the reactance of the series capacitances is very high and the reactance of the parallel capacitances is very low, resulting in the equivalent circuit depicted in Figure 16-4(d).

The mid-frequency coupling gain of the circuit in Figure 16-4(e) is:

$$A_V = \frac{V_L}{V_{oc}} = \frac{R_L}{R_{int} + R_L} \tag{16.9}$$

or

$$Av\ dB = 20\log A_V$$

The low-frequency cutoff of the amplifier is at the point the reactance of the equivalent low-frequency capacitances equals the equivalent resistance or where $X_C = R$ and the phase angle $\theta = -45°$.

$$f_{co} = \frac{1}{2\pi RC} \tag{16.10}$$

Example 12   Find the low-frequency cutoff for an amplifier when $C = 1\mu F$ and $R = 1.5\ k\Omega$.

$$f_{co} = \frac{1}{2\pi \times 1.5\ k \times 1\mu F} = \frac{0.159}{1.5 \times 10^{-3}} \approx 100\ Hz$$

The low frequency response is shown in Figure 16-2(a). The cut-off frequency point is defined as: $0.707\ E_{maximum}$, $0.707\ I_{maximum}$, $-45°$ and one-half maximum power.

The high-frequency response is calculated using the shunt capacitances in parallel with the load and the internal resistances.

Example 13   Find the high frequency response of amplifier in Figure 16-3 when $C_S = 100\ pF$ and R_{load} in parallel with $R_{int} = 10\ k\Omega$.

$$f_{co} = \frac{1}{2\pi RC}$$

The internal resistance is $10\ k\Omega$ and the shunt capacitance is $100\ pF$. The cutoff frequency is:

$$f_{co} = \frac{1}{2\pi RC} = \frac{0.159}{1 \times 10^4 \times 1 \times 10^{-10}}$$

$$f_{co} = 0.159 \times 10^6 = 159\ kHz$$

The Bode plot for the filter is shown in Figure 16-2. Again, the plot is constructed by:

1.   Marking the cutoff frequency at the 0 dB point on the semi-log paper.
2.   Locating the frequency point of either -6 b at one octave or -20 dB at one decade.
3.   Drawing a straight line between these points.

A simple RC or RL (single pole) will have a roll-off of -6 dB per octave and -20 dB per decade.
    The band-pass filter is a combination of the high- and low-pass filters and is plotted as shown in Figure 16-2.
    We note from Figure 16-5 that the gain of the amplifier is flat from the low-cutoff frequency to the high-cutoff frequency resulting in a *band pass*. The band pass is calculated:

$$\text{Band pass} = f_h - f_l$$
$$\text{Band pass} = 159\ kHz - 10.6\ Hz \approx 159\ kHz \tag{16.11}$$

The center of frequency of the band pass is the *logarithmic center* of the band pass:

$$\text{Center frequency} = \sqrt{f_h \times f_l} \tag{16.12}$$

Example 14 Calculate the values of f_1, f_2, f_{mid}, and plot the bandwidth on semi-log paper using the equivalent circuit shown in Figure 16-4; when $R_{int} = 5$ kΩ, $C_C = 5$ μF, $R_L = 5$ kΩ, and $C_S = 100$ pF.

Mid-frequency gain

$$A_{mid} = \frac{R_L}{R_L + R_{int}} = \frac{5\,k}{5\,k + 5\,k} = \frac{5\,k}{10\,k} = 0.5$$

$$A_V\ dB = 20\log 0.5 = 20(-0.3010) = -6$$

Low-frequency (c) yields

$$R = R_{int} + R_L = 10\ k\Omega$$

$$C = C_C = 0.5\ \mu F$$

$$f_1 = \frac{1}{2\pi RC} = \frac{1}{6.28 \times 10\,k \times 0.5\ \mu F}$$

$$= \frac{.159}{5 \times 10^{-3}} = \frac{159}{5} \approx 32\ \text{Hz}$$

High-frequency circuit yields

$$R_T = R_{int}\ ||\ R_L = 5\,k\ ||\ 5\,k = 2.5\ k\Omega$$

$$f_2 = \frac{1}{2\pi RC} = \frac{0.159}{2.5 \times 10^3 \times 100 \times 10^{-12}} = \frac{0.159}{.25 \times 10^{-6}} \approx 636\ \text{kHz}$$

Bandwidth equals:

$$f_h - f_L = 636\ \text{kHz} - 2.5\ \text{kHz} \approx 633.5\ \text{kHz}$$

The mid-frequency:

$$f_{mid} = \sqrt{f_1 f_2}$$

Exercises 16–3 Find the cutoff frequencies, the mid-frequency gain, the mid-frequency of a transistor amplifier, and draw the Bode plot for each circuit.

1. When $R_1 = 5$ kΩ, $R_2 = 10$ kΩ, $C_1 = 5$ μF, and $C_2 = 1$ nF.
2. When $R_1 = 10$ kΩ, $R_2 = 8$ kΩ, $C_1 = 5$ μF, and $C_2 = 10$ nF.
3. When $R_1 = 1$ kΩ, $R_2 = 4$ kΩ, $C_1 = 5$ μF, and $C_2 = 100$ pF.
4. When $R_1 = 1$ kΩ, $R_2 = 10$ kΩ, $C_1 = 2$ μF, and $C_2 = 100$ pF.
5. When $R_1 = 2$ kΩ, $R_2 = 12$ kΩ, $C_1 = 1$ μF, and $C_2 = 159$ pF.

Problems 16–2

1. A radio receiver with an input of 100 mW produces 3 W into the speaker. What is the overall decibel gain of the receiver?
2. The input stage of a preamplifier produces 200 mW output with a 50 micro watt input. What is the overall dB gain of the amplifier?
3. A transmission line with an 8.7 watt input dissipates 1.2 watts along the impedance of its length. What is the decibel loss of the line?
4. What power level is represented by a 33 decibel gain, with a zero reference level of 600 mW?

5. A dBm measurement gives a reading of ± 15 dB. What is the power level represented?
6. A microphone with an output level of -65 dB is connected to a preamplifier both with a gain of $+56$ dB which is connected through an attenuator pad with a loss of -8 dB and into a power amplifier with a gain of 70 dB. What is the total dB gain of the system and what is the output power if the output of the microphone is 1 mW?
7. If the output of the power amplifier in Problem 6 is 100 watts, what is the input power levels: at the preamplifier and at the power amplifier?
8. Suppose the output of the power amplifier in Problem 6 is 30 volts, what is the output voltage at the microphone if the dB gain of the system is $+100$ dB? Assume that the loads of the amplifier and the microphone are equal. *Note:* dB $= 20 \log V_o / V_{in}$.
9. A two-stage transistor amplifier has a 26 decibel first-stage gain, a 31 decibel second-stage gain, and the coupling circuit to the load develops a -3 decibel loss. What is the overall gain of the amplifier?
10. With a 25 nW input, what is the output power of the amplifier in Problem 9?

Summary

1. The logarithm of a number is the power of a certain base that equals the number.
2. Common logarithms utilize the base 10.
3. Natural logarithms utilize the base ϵ, an irrational number of approximately 2.302585093....
4. Applications of both natural and common log in electrical problems are best solved by the use of a scientific calculator.

Chapter 17

Natural Logarithms and Exponential Functions

▶ ## 17.1 Introduction

Logarithms based on the irrational number 2.71828..., or ϵ (epsilon) are called *natural logarithms*, or *Naperian logarithms*, after John Napier, who invented the logarithmic method of calculation. The natural logarithm of a number is:

$$\epsilon^x = N$$

or

$$x = \log_\epsilon \tag{17.1}$$

where $\epsilon = 2.71828\ldots$.

It would be difficult to overstate the importance of natural logarithms and exponential equations [Exponential equations are equations in which the unknown appears in the exponent, e.g. ($15 = 2^x$)] in the mathematics of electricity and electronics. For example, when a capacitor discharges through a resistance, the plot of current versus time follows an exponential law that is often called the natural law of decay. Figure 17-1 illustrates this exponential fall of capacitor voltage. The formula for the curve in Figure 17-1(b) is

$$v = E\epsilon^{-t/RC} \tag{17.2}$$

where v is the voltage across the capacitor at any time t in seconds after the switch is closed, E is the voltage across the capacitor the instant the switch is closed, ϵ equals 2.71828, R is the resistance value in ohms, and C is the capacitance value in farads.

If the initial voltage across the capacitor terminals is stipulated as 100 percent, the formula for the percentage of v or the instantaneous value of v becomes

$$v = \epsilon^{-t/RC} \tag{17.3}$$

Note that the exponent is negative in the foregoing exponential equation. In turn, v becomes smaller as larger values of t are assigned. Observe also that when t equals RC seconds, the equation becomes

$$v = \epsilon^{-1} = \frac{1}{2.71828} = 0.368\, RC \tag{17.4}$$

In any RC circuit (Figure 17.1(a)), the terminal voltage of the capacitor will decay to 36.8 percent of its initial value at the end of RC seconds, as noted in Figure 17-1(b). Since the vertical axis denotes percentage values, the horizontal axis denotes time in RC units; this is a *universal* or generalized chart that applies to any series RC circuit. This chart is called a *universal time-constant chart*.

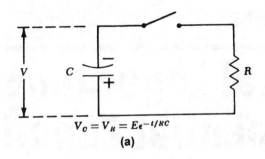

$$V_0 = V_R = E\epsilon^{-t/RC}$$

(a)

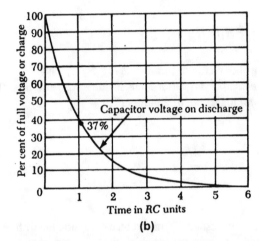

Capacitor voltage on discharge

37%

Time in RC units

(b)

Figure 17–1

Example of an exponential function: (a) Capacitor with voltage E about to be discharged. (b) Decay of capacitor voltage versus time in RC units.

In turn, the product RC is called the *time constant* of an RC circuit, which is the time in seconds required for the terminal voltage of the capacitor to discharge 63.2 percent to 36.8 percent of its initial value after the switch is closed as in Figure 17-1(a).

As would be anticipated, an exponential formula (17.5) also describes the charging of a capacitor through a resistance from a voltage source.

$$V_c = 1 - \epsilon^{-t/RC} \tag{17.5}$$

Figure 17-2 is a universal time-constant chart for charge and discharge of a capacitor through a resistor. The rise of voltage across the capacitor while it is being charged is expressed by curve A, and the decrease of circuit current and voltage across the resistor are expressed by curve B. For the purpose of plotting a universal RC time-constant chart, we let E equal 1, or 100 percent. Thus, the formula of the charging curve in Figure 17-2 is written

$$V = 1 - \epsilon^{-t/RC} \tag{17.6}$$

Since $\epsilon^{-t/RC} = 1 - V$, we can also write

$$\frac{-t}{RC} = \log_\epsilon(1 - V)$$

Summary of Formulas For the discharge of the capacitor,

$$I = \frac{E}{R}\epsilon^{-t/RC}$$

$$V_c = V_R = E\epsilon^{-t/RC} \tag{17.7}$$

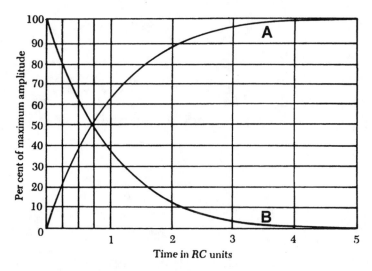

Figure 17-2 Universal *RC* time-constant chart.

For the charging of a capacitor,

$$I = \frac{E}{R}\epsilon^{-t/RC}$$

$$V_c = E(1 - \epsilon^{-t/RC})$$

$$V_R = E\epsilon^{-t/RC} \tag{17.8}$$

 ## 17.2 Natural Logarithms

As with common logs, the scientific calculator is the most advisable method of finding the value of natural logarithms.

For example, to find the natural log of 4, enter 4 into the calculator and press LN. To find the antilog, enter the antilog into the calculator and pressing 2nd or INV and LN.

Solution of exponential equations such as $V = \epsilon^{-t/RC}$ is facilitated by the use of exponential functions on a scientific calculator. For example, ϵ^3 is found by entering 3 into the calculator and press 2nd and LN to yield 20.085.

Example 1 Determine the time t when the voltage across the capacitor in the circuit in Figure 17-1(a) will reach 75 V. $R = 100$ kΩ, $C = 100$ μF, $E = 100$ V, and $V_c = E(1 - \epsilon^{-t/RC})$.

Substituting,

$$75 = 100(1 - \epsilon^{-t/RC})$$

$$\frac{75}{100} - 1 = -\epsilon^{t/RC}$$

taking the ln of each side,

$$.75 - 1 = -\epsilon^{-t/RC}$$

$$-.25 = -\epsilon^{-t/RC}$$

Log each side

$$\ln 0.25 = -\frac{t}{RC} \ln \epsilon$$

$$-1.386 = -\frac{t}{RC}(1)$$

$$t = 1.386 \times RC$$

Then

$$-1.386 = -\frac{t}{RC}$$

$$t = 1.386 \, RC$$

Finally,

$$t = 1.386 \times 100 \times 10^3 \times 100 \times 10^{-6}$$

$$t = 13.86 \text{ seconds}$$

Problems 17–1

1. From the formula $N = (1 + 1/m)^m$, determine the sum of N for m values of 1, 2, 3, 4, and 5.
2. The time (t) of a chemical reaction used to etch printed circuits depends upon the Centigrade temperature (T) according to the formula $t = 1.2^{0.18T}$. Find t when $T = 30°C$, then $T = 55°C$, and when $T = 150°C$.
3. Determine the time in seconds for the voltage across the resistor in Figure 17-1(a) to reach 10 V when $R = 10$ kΩ, $C = 100$ pF, and $E = 100$ V.
4. Determine the time in seconds for the current to reach 2.2 mA in Problem 3.
5. What percentage of a time-constant (RC) is required for a capacitor in an RC circuit to charge to 10 percent of the applied voltage?
6. What percentage of a time-constant (RC) is required for a capacitor in an RC circuit to charge to 90 percent of the applied voltage?
7. What number of time-constants occur between the 10 percent charge in Problem 5 and the 90 percent charge in Problem 6?
8. Determine the charge time of the capacitor in the relaxation oscillator when the diode switches on at 20 volts and switches off at 0.5 volts. The resistance of the diode when it is off can be considered to be infinity.
9. What is the frequency of the oscillations in Problem 8 where $f = 1/t$?
10. The rise time of the pulse in Problem 8 is measured as the time duration between $V_c = 10$ percent and $V_c = 90$ percent. What is the rise time?
11. The fall time of the pulse in Problem 8 is measured as the time duration between $V_c = 90$ percent, and $V_c = 10$ percent. What is the fall time?

▶ 17.3 Inductance and Exponential Formulas

Exponential formulas are also encountered in the analysis of inductive circuits. When a voltage is applied to an inductor a counter electromotive force (cemf) is developed that acts to limit the increase of current. This is an analogous situation to the storage of energy in the electrostatic field of a charged capacitor.

The following conditions apply at the first instant when a voltage of ten volts is applied to an *LR* circuit,

The voltage across the inductor L, due to cemf, is equal to ten volts,
Circuit current is zero,
The voltage across the resistor is zero.

Voltage across the inductor begins to decrease with time in accordance with curve B of the universal time-constant chart in Figure 17-2. Evidently, the current and voltage drop across *R* rises in accordance with curve A.

Note that the horizontal axis in Figure 17-2 can be marked off in *L/R* units. In other words, the time-constant of an *LR* circuit is equal to *L/R* seconds, just as the time-constant of an *RC* circuit is equal to *RC* seconds.

In one time-constant the conditions are:

$$V_L = 63.2\% \; E$$

$$I = 63.2\% \; I \text{ maximum}$$

$$V_L = 36.8\% \; E$$

In 5 times-constants, we say that the circuit has stabilized, then

$$I = \frac{E}{R}$$

$$V_R = E$$

$$V_L = 0$$

If sufficient time passes for the circuit to stabilize and the input voltage is suddenly reduced to zero, the magnetic field starts to collapse in *L*. At the first instant, the cemf is equal to *E*, and this cemf has a polarity which causes current flow to continue in the same direction as before, in accordance with the law of conservation of energy. The cemf is applied across *R*, and the stored energy in the magnetic field is progressively dissipated as I^2R loss in *R*. The cemf falls and current flow decreases in accordance with curve B. At the end of five time-constants, the stored energy in the magnetic field has decreased to practically zero.

As would be anticipated, curves A and B in Figure 17-2 have exponential formulations that are similar to the formulations for an *RC* series circuit. Let us summarize the formulas for an *LR* circuit when a d-c voltage is applied. The formula for curve B in Figure 17-2 is written

$$v_L = E\epsilon^{Rt/L} \tag{17.9}$$

where *v* is in volts, epsilon equals 2.718, *R* is in ohms, *t* is in seconds, and *L* is in henrys.

The formula for curve A in Figure 17-2 is written

$$i = 1 - \epsilon^{-Rt/L} \tag{17.10}$$

The time constant formulas for both *RC* and *LR* circuits follow a constant pattern. For example, the exponential function in each formula is:

$$\epsilon^{-t/time \; constant}$$

The time constant for a capacitor and resistor is *RC*, therefore:

$$\epsilon^{-t/RC}$$

The time constant for an inductor and a capacitor is L/R, therefore:

$$\epsilon^{-t/L/R} = \epsilon_{-tR/L}$$

$$e_R = E\epsilon^{-Rt/L} \qquad\qquad (17.11)$$

Problems 17–2

1. What is the time-constant of a 10 mH inductance connected in series with a 1 MΩ resistor?
2. What value of a capacitor will give a time-constant of 100 μsec when connected in series with a 1 MΩ resistor?
3. What is the time-constant of a 200 mH inductor in series with a 1 kΩ resistor?
4. Suppose a 100 V pulse is applied across the LR circuit in Problem 3. What length of time must pass for the voltage across the resistor to rise to 87 V? What is the energy stored in the inductor at that instant?
5. A 10 H inductor and a 100 Ω resistor connected in series are placed across a 200 V d-c supply. What current flows in the circuit 20 msec later?
6. What current will flow in the circuit described in Problem 5 in 100 msec? What energy is stored in the inductor during this time?
7. A steady current of 10 amp flows through a 100 H inductor.
   (a) What voltage would be produced if a 100 kΩ resistor could be placed across the inductor instantaneously?
   (b) What energy is stored in the inductor?
   (c) What energy would the resistor dissipate at the first instant?
   (d) What current would flow after 10 μsec, and what amount of energy would the inductor still store?
8. A steady current of 1 A flows in a series LR circuit. What L/R value would be required to maintain a current of 100 mA of 100 μsec after the applied voltage is reduced to zero?
9. If $\epsilon^{-x} = 0.3678\ldots$, what is the value of x?
10. What is the time constant of a 0.001 μF capacitor connected in series with a 100 Ω resistor?

Summary

1. Natural logarithms are based on the number $2.71828\ldots$, an irrational number.
2. The universal time-constant chart is a graph of the current and voltage action in either an RC circuit or an LR circuit when a constant input voltage is applied.
3. The universal time-constant chart can be utilized to obtain approximate waveform results in RC or LR circuits when a square wave is applied to the circuits.

Chapter 18

Fundamentals of Trigonometry

 18.1   Angular Notation

A simple angle is formed when two *straight* lines meet. We may define an angle as the difference in direction of two straight lines. The lines are called the *sides*, and their point of intersection is called the *vertex* of the angle. The size of an angle depends upon the amount of divergence of the sides, and the size of an angle is independent of the lengths of its sides. As depicted in Figure 18-1, if one straight line intersects another straight line perpendicularly, the angles that are formed are equal, and the lines intersect at *right angles*. An *acute angle* is less than a right angle. An *obtuse angle* is greater than a right angle.

In the most general sense, an angle is generated by revolution of a line, as shown in Figure 18-2. Note that there is no limit to the number of times that line may revolve about point 0, and therefore there is no limit to the size of an angle that may be generated. With reference to Figure 18-2, we call point 0 the *origin*, and we call the revolving line the *radius vector*. Line 0A is called the *initial side*, and line 0B is called the *terminal side* of the angle. By definition, the angle A that is generated is said to be *positive* if the radius vector rotates *counterclockwise*; on the other hand, the angle is said to be *negative* if the radius vector rotates *clockwise*.

Three systems of angular measure are in common use. Mechanics commonly use the *sexagesimal system*; technicians and engineers generally use *radian* measure; the armed services ordinarily use *mil* measure. In sexagesimal measure, one complete revolution of the radius vector is stipulated to encompass 360 degrees (360°), and the degree is thereby defined as the unit of angular measure. Thus, 1/360 of a complete revolution is equal to 1°. The degree may be divided in minutes or seconds.

Next, let us consider radian measure. A *radian* (Figure 18-3) is defined as the angle subtended by an arc of a circle equal to the length of the radius of the circle. In other words, we visualize that

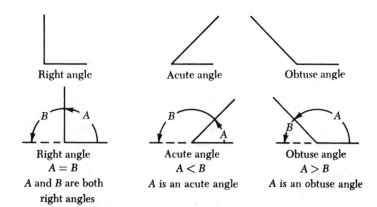

Figure 18–1
Three basic types of angles.

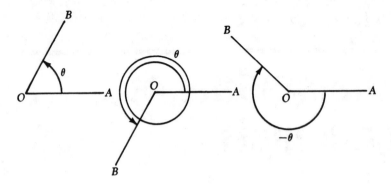

Figure 18–2 Generation of angles by revolution of a line.

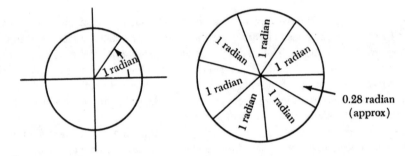

Figure 18–3 Visualization of radian measure.

we have taken the radius of the circle and have "bent" the radius to lie along the circumference of the circle; the ends of the radius that we have thus "bent" define an angle of 1 radian. We know that the circumference of a circle has length equal to 2π times the length of half its diameter, or $2\pi r$, where r is the radius of the circle. Accordingly, the ratio of circumference to radius is equal to $2\pi r$. This is just another way of saying that a circle contains 2π radians of angular measure. Thus,

$$2\pi \text{ radians} = 360°$$

or

$$1 \text{ radian} = 57.3°, \text{ approximately} \tag{18.1}$$

1. *To change radians to degrees, multiply the number of radians by 57.3°; the product is an approximation to the number of degrees in the angle.*
2. *To change degrees to radians, multiply the number of degrees by 0.01745 radian; the product is an approximation to the number of radians in the angle.*

For maximum accuracy in application of these rules, use the associated formulas with π expressed to as many decimal places as required.

3. *To change from degrees to radians, multiply the number of degrees by $\pi/180$.*
4. *To change from radians to degrees, multiply the number of radians by $180/\pi$.*

Example 1   Change 2.8 radians to degrees.

$$2.8 \text{ radians} \times 57.3°/\text{radians} = 162.44°$$

Exercises 18–1 Change the following angles in degrees to radians:

1. 27.5°   2. 186.2°   3. 75.3°   4. 256°

5. 315°   6. 60°   7. 345.4°   8. 17.2°

Change the following angles in radians to angles in degrees:

9. 2π rad   10. 4 rad   11. 2.4 rad   12. 3.6 rad

13. 0.5 rad   14. 1.8 rad   15. 5.6 rad   16. 7.2 rad

▶ 18.2 Trigonometric Functions

It was previously noted that trigonometric functions are certain ratios that depend upon angles, or trigonometric functions are functions of an angle. We will begin by defining the trigonometric functions of acute angles in terms of the sides of a *right triangle*. If we draw a triangle that has one *right angle*, we have drawn a right triangle, as shown in Figure 18-4. A right angle is an angle of 90°. The *altitude* of the right triangle is denoted by a; the *base* of the right triangle is denoted by b; the *hypotenuse* of the right triangle is denoted by c.

The *sum of the angles of any triangle is equal to 180°*. Accordingly, a right triangle contains one right angle and two acute angles. Furthermore, the sum of these acute angles must be equal to 90°. In turn, we may write

$$\text{angle } \theta + \text{angle } \phi + \text{angle } 90° = 180° \tag{18.2}$$

$$\text{angle } \theta + \text{angle } \phi = 90° \tag{18.3}$$

You have perhaps already learned in a geometry course that the Pythagorean theorem states: *The square of the hypotenuse is equal to the sum of the squares of the other two sides of a right triangle.* Thus, with reference to Figure 18-4, we may write

$$h^2 = o^2 + a^2 \tag{18.4}$$

$$h = \sqrt{o^2 + a^2} \tag{18.5}$$

Figure 18–4
The nomenclature of a right triangle; the acute angles of the right triangle are denoted by θ and ϕ.

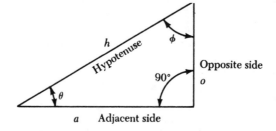

Evidently, we may also write

$$h = \sqrt{o^2 + a^2} \tag{18.6}$$

$$a = \sqrt{h^2 - o^2} \tag{18.7}$$

There are six trigonometric functions of any angle, however, we will confine our study to the three most often used in electronics.

Sine θ = opposite side/hypotenuse = o/h.
Cosine θ = adjacent side/hypotenuse = a/h.
Tangent θ = opposite side/adjacent side = o/a.

We might note the functions above could also be written for angle ϕ.

 ## 18.3 Trigonometric Functions of 30°, 45°, and 60°

Next, let us calculate the values for the trigonometric functions of some particular angles. With reference to Figure 18-4, we will calculate these functions for an angle of 45°. The base and altitude are one unit in length each, and it follows from the Pythagorean theorem that the hypotenuse has a length of two units. Both acute angles have a value of 45°. In turn, we write

$$\sin 45° = \frac{o}{h} = \frac{1}{\sqrt{2}} = 0.707$$

$$\cos 45° = \frac{a}{h} = \frac{1}{\sqrt{2}} = 0.707$$

$$\tan 45° = \frac{o}{a} = \frac{1}{1} = 1 \qquad \text{(18.8)}$$

Next, with reference to Figure 18-5, let us calculate the values of the trigonometric functions for an angle of 60°. This is an equilateral triangle, and the length of each side is two units. Each angle has a value of 60°. The bisector of an angle is the perpendicular bisector of the opposite side, resulting in two 30° angles. In turn, we may write

$$\sin 30° = \frac{o}{h} = \frac{1}{2} = 0.5 \qquad \text{(18.9)}$$

$$\cos 30° = \frac{a}{h} = \frac{\sqrt{3}}{3} \approx 0.866 \qquad \text{(18.10)}$$

$$\tan 30° = \frac{o}{a} = \frac{1}{\sqrt{3}} \approx 0.577 \qquad \text{(18.11)}$$

Similarly, we may write

$$\sin 60° = \frac{o}{h} = \frac{\sqrt{3}}{2} \approx 0.866 \qquad \text{(18.12)}$$

$$\cos 60° = \frac{a}{h} = \frac{1}{2} = 0.5 \qquad \text{(18.13)}$$

$$\tan 60° = \frac{o}{a} = \frac{\sqrt{3}}{1} \approx 1.732 \qquad \text{(18.14)}$$

Sine, Cosine, or Tangent are *looked* up by placing the degrees of the angle into the calculator and pressing the proper button. For example, entering 34.5 into the calculator and pressing the Sine button yields Sine 34.5° = 0.5664062. In most technical work the number would be rounded off to 0.566.

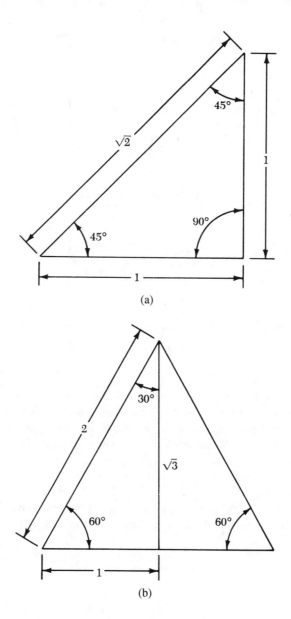

Figure 18-5
Configurations of a triangle: (a) an isosceles right triangle, (b) a 30°, 60° right triangle.

(a)

(b)

 ## 18.4   *General Angles*

It is necessary in engineering work to find trigonometric functions of any angle; otherwise stated, we need to know how to find the trigonometric functions of 100°, 275°, and so on.

The two numbered scales which intersect perpendicularly at their origins in Figure 18-6, form a *rectangular coordinate* system. The scales are called the axes, denoted OX and OY. The point 0 is called the *origin*. If P is any point in the plane of OX and OY, the *horizontal coordinate* of point P is the perpendicular distance x from P to OY. This distance is *positive* when P is to the right of OY, but *negative* when P is to the left of OY. The horizontal coordinate is called the *abscissa*. Again, the vertical coordinate of point P is the perpendicular distance y from P to OX. This distance is *positive* when P is *above OX*, but *negative* when P is *below OX*. The vertical coordinate is called the *ordinate*.

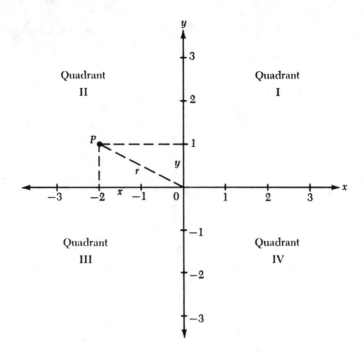

Figure 18–6
A rectangular coordinate system.

Together, the abscissa and ordinate are called the *coordinates* of point P, and are denoted by (x, y). The distance OP is called the *radius vector*, and OP is always positive, except when $OP = 0$. The coordinate axes separate the plane into four *quadrants*, which are numbered *counterclockwise* I, II, III, IV. Any angle placed so that 0 is located at the origin of the coordinate system, and with its initial side along the positive X axis is said to be in *standard position*. Three angles in standard position are shown in Figure 18-7. Their values are $\theta = 30°$, $\phi = 150°$, and $\gamma = -60°$.

Note that x and y are both positive in the first quadrant, and accordingly, all the functions are positive in the first quadrant. However, x is negative and y is positive in the second quadrant;

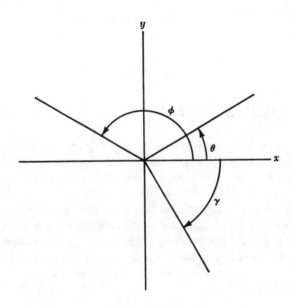

Figure 18–7
Three angles in standard position.

in turn, all the functions are negative except sin θ. In the third quadrant the sine and cosine are negative. Finally, in the fourth quadrant the tangent is negative.

Example 2   Find the trigonometric functions of 210°.

$$\sin 210° = -\sin 30° = -0.5$$
$$\cos 210° = -\cos 30° = -0.866$$
$$\tan 210° = \tan 30° = 0.577$$

Note that the angle of 210° forms a 30° angle to the x axis

$$210° - 180° = 30°$$

in the third quadrant.

Exercises 18–2   Find the trigonometric functions for sine, cosine, and tangent of the following:

1.	60°	**2.**	30°	**3.**	45°	**4.**	135°
5.	150°	**6.**	120°	**7.**	225°	**8.**	315°
9.	−120°	**10.**	−30°	**11.**	−300°	**12.**	780°

Summary

1. The radius of a circle is the number of degrees that would be covered if the radius of that circle were bent along the circumference.
2. A radian of a circle is approximately 57.3°.
3. The right triangle is utilized to solve values of current, voltage, and phase angle in RC, LR, and LCR circuit voltage and phase angle in RC, LR, and LCR circuits when an AC voltage is applied.
4. Application of the sine, cosine, and tangent are necessary to solve for current, voltage, and impedance values in AC circuits.
5. The Pythagorean theorem is used to determine the relationship between applied voltage and voltage drops in RL and RC circuits.
6. The Pythagorean theorem is used to determine the relationship between circuit impedance (Z) and circuit reactance (RL or RC) circuits.

Chapter 19

Vector Fundamentals

 ## 19.1 Introduction

A vector value is a quantity that is specified by a magnitude, a direction, and a sense. Thus, "north-and-south" is a direction, but a compass needle also senses "north." In geometric form, a vector is represented by a line segment, the *length* of which denotes the *magnitude* of the vector, the *orientation* of which denotes the *direction* of the vector, with the *sense* of the vector indicated by an *arrowhead*. Recall that a *radius vector* rotates counterclockwise, and is associated with a sine wave. The magnitude of this vector remains constant, and its sign remains constant; on the other hand, the orientation of the vector is changing progressively in accordance with its angular velocity.

Note that at zero time, the radius vector is directed to the right along the time axis. At time t in Figure 19-1, the radius vector has rotated through the angle θ, and $R \sin 0$ is equal to the length of the line Pt. Since the *vectorial angle* θ represents a phase, the radius vector is sometimes called a *phasor*.

 ## 19.2 Periodic Functions

Since the graph of a trigonometric function repeats itself over each interval of 2π radians, we call the function a *periodic function*. Recall that we discussed the angular velocity of a point moving counterclockwise around a circle. We express angular velocity in radians per second. We write

$$\omega = 2\pi f \text{ radians per second} \qquad (19.1)$$

where the radius vector is rotating f revolutions per second.

Equation 19.1 states that if the radius vector in Figure 19-1 is rotating one revolution per second, it sweeps over 2π radians in one second. The projection of the tip of the radius vector to the horizontal axis is equal to $\sin \theta$, since this is a unit circle. In turn, if we plot its projection versus time, we will plot a sine curve, or sine wave.

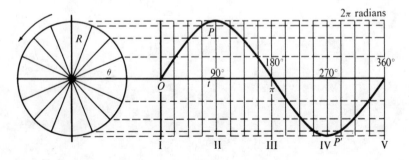

Figure 19–1 Generation of a sine curve from a projection of the radius vector.

The total angle θ generated by the rotating radius vector is

$$\theta = \omega f \text{ radians} \tag{19.2}$$

Example 1   The radius vector r in Figure 19-1 rotates at a rate of 2,000 revolutions per second. What is the angular rotation (θ) in radians after 1 msec?

$$\theta = \omega t$$

$$= 2\pi \times 2000 \times 0.001$$

$$= 4\pi \text{ radians}$$

If we denote the length of the projection of the radius vector as y, and denote time by x, we write

$$y = \sin x \tag{19.3}$$

or

$$y = 360° \text{ ft} \tag{19.4}$$

We call the value of y the amplitude of the sine wave at any instant in time. The number of revolutions that the radius vector makes per second is called its *frequency*. Each time the radius vector makes one complete revolution it has completed one *cycle*. Frequency is measured in Hertz (Hz). One Hertz is equal to one cycle per second. It is evident that

$$f = \frac{\omega}{2\pi} \text{ Hz} \tag{19.5}$$

The reciprocal of frequency is called the *period* of the sine wave. Thus,

$$T = \frac{1}{f} \text{ sec} \tag{19.6}$$

Example 2   A point moves in a circular path at a velocity of 100π radians per second.

(a)   What is its angular velocity in degrees per second?
(b)   What is the period of 1 cycle?
(c)   What is the frequency of repetition?

Solution:

(a) 100π radians/sec $\times$ 360°$/2\pi$ radians $= 18{,}000°$ sec

(b) $f = \dfrac{\text{degrees/sec}}{360°/\text{cycle}} = \dfrac{18{,}000}{360} = 50 \text{ Hz}$

(c) $T = 1/f = 1/50 = 0.02$ sec

The two sine waves shown in Figure 19-2 both have the same frequency. However, they cross the x axis so that their radius vectors are separated by a constant angle θ. The two sine waves differ in phase by $\pi/6$ radians. This angle is called the *phase angle*. Since waveform A crosses the horizontal axis before waveform B, we say that A leads B. Conversely, we say that B lags A.

In Figure 19-2(b), waveform B crosses the horizontal axis before waveform A and we say that B leads A or that A lags B.

The waveforms may be formulated

$$y = r \sin(\omega t \pm \theta) \tag{19.7}$$

where θ is the phase angle.

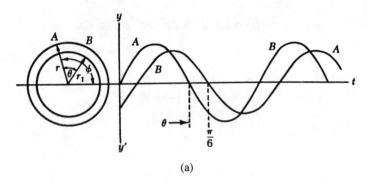

(a)

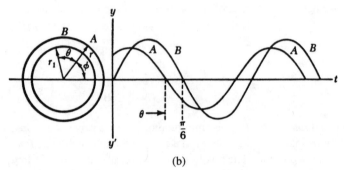

Figure 19–2
Representation of phase
differences: (a) A leads B
by $\pi/6$ radians,
$B = r_1 \sin(\omega t - \theta)$, (b) A
lags B by $\pi/6$ radians,
$A = r_1 \sin(\omega t - \theta)$.

(b)

Exercises 19–1 On the same axis, plot the following equations for 1 cycle.

1. $y = \sin \omega t$
2. $y = \cos \omega t$
3. $y = 2 \sin \omega t$
4. $y = 3 \sin \omega t$
5. $y = 2 \cos \omega t$

In the following equations, find the amplitude, frequency, period, velocity, and phase angle.

6. $y = 12 \sin(377t + \pi/3)$
7. $y = 220 \sin(628t - 3/2\pi)$
8. $y = 100 \sin(2500t - \pi/6)$
9. $y = 25 \sin(120.5t + 2\pi/3)$
10. $y = 300 \sin(628t - \pi/4)$

 19.3 Vector Notations

Vectors are also commonly symbolized by indicating their magnitudes and angles. For example, we might write $15\angle 30°$. The number 15 might represent pounds, miles per hour, or other concrete numbers. If we are considering two vectors, we might symbolize the vectors in literal notation by writing $0A\angle\theta$, and $0B\angle\theta$. These expressions tell us that the vectors have different *magnitudes*, but that they have the *same phase*. In numerical notation, we might write $15\angle 30°$ and $25\angle 30°$.

▶ 19.4 Applications of Vectors

Suppose that we have a vector Z, as depicted in Figure 19-3, and we drop a perpendicular from the tip of the vector to the horizontal axis; we obtain the component vectors X and R, which are called the *triangular components* of Z. In other words, vectors X and R are at right angles to each other. We call the length of X the *vertical component* of vector Z, and we call the length of R the *horizontal component* of vector Z. *Triangular* components play an important role in calculations of circuit action. Observe that $X = Z \sin \theta$, and that $R = Z \cos \theta$, or $Z = X \sin \theta = R \cos \theta$. In these expressions, X, R, and Z denote the magnitudes of the vectors. It is also evident that:

$$Z^2 = R^2 + X^2$$

This expression for vector Z denotes its magnitude, but not its angle. To completely describe the vector Z, we may write:

$$Z = \sqrt{R^2 + X^2} \ \angle \theta \quad \text{or} \quad Z \angle \theta$$

This equation expresses the vector in polar form.

Example 3   Convert the vector whose rectangular parts are $R = 300$ and $X = 400$ to its equivalent polar form, using Figure 19-3.

$$\tan \theta = \frac{X}{R} = \frac{400}{300} \approx 1.33$$

$$\theta = 53.2°$$

$$|Z| = \frac{X}{\sin \theta} \ \text{or} \ \frac{R}{\cos \theta}$$

$$|Z| = \frac{X}{\sin 53.2°} = \frac{400}{0.8} = 500$$

It follows from the foregoing discussion that it is very important for us to *visualize* what we are doing when we make vector calculations. If we have a clear picture of the problem, we will not be in doubt whether we should add, subtract, calculate a square, or extract a square root, and so on. In fact, if a neat and accurate diagram is drawn, a *graphical solution* will often be sufficiently accurate for routine work. When better accuracy is required, we calculate the magnitude and

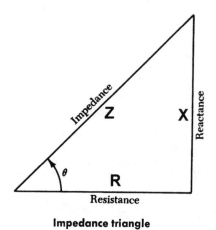

Figure 19-3
Representation of a triangular form of impedance.

Impedance triangle

angle of the resultant with the aid of a calculator, and merely use the diagram as a "map" of the course to be followed in calculation.

Thus, far, we have considered only the addition and subtraction of vectors. However, as we proceed with analysis of circuit action, we will often be called upon to multiply and divide vectors. Some circuit problems will require that we raise quantity to a power, or take a root of a vector. In these operations, we will be concerned with operating on the *angle* of the vector as well as operating on its *magnitude*. For example, we will learn how to multiply $35\angle15°$ by $13\angle25°$, how to divide $35\angle15°$ by $13\angle25°$, and so on. We will find that these new operations are logical extensions of the basic facts that have been explained in this chapter.

Exercises 19–2 Convert the following forces in polar form to rectangular form:

1. $100\angle60°$
2. $20\angle30°$
3. $100\angle90°$
4. $20\angle45°$

5. $100\angle0°$
6. $25\angle53.2°$
7. $100\angle38.8°$
8. $75\angle25°$

9. $77\angle135°$
10. $180\angle7°$
11. $80\angle55°$
12. $12.700\angle300°$

Problems 19–1

1. A jet aircraft flying 600 mph ground speed due east runs into a crosswind of 60 mph blowing north. What is the plane's new direction and speed before the pilot corrects his course?
2. What new course and speed is necessary in Problem 1 for the aircraft to maintain its original ground speed and direction?
3. A sail boat is being propelled northeast by a 20 knot wind. What is the resultant speed and direction if the current is southeast at 4 knots?
4. A missile is fired at an angle of 83°. If its average speed is 1200 mph, when will the missile be 150 miles above the earth? (Assume the earth to be a flat surface.)
5. At the time solved for in Problem 4, would the missile be more or less than 250 miles above the earth?
6. A submarine rises at a vertical rate of 120 feet per sec. If the vessel surfaces in 5 sec, during which time the horizontal distance covered is found to be 1,000 feet, what angle will be formed with the surface of the water as the bow comes out of the water?
7. Two forces, 150 lb and 200 lb, 30° apart, act on a body. What resultant force would give the same magnitude of force and direction?

Summary

1. One cycle of a sine wave represents 360° or 2π radians.
2. The frequency of a waveform is in cycles per second called Hertz.
3. The period of a waveform is the reciprocal of the frequency.

Chapter 20

Principles of Alternating Currents

▶ 20.1 Introduction

Previous discussion has been concerned with either steady flow of current and constant voltage (d-c), or unidirectional current flow that varies in amplitude (pulsating d-c). Now, we will consider current flow that changes its *direction* periodically; this type of electric current is called *alternating current* (a-c). It is accompanied by an alternating voltage which changes its polarity periodically. We commonly write "a-c voltage" to denote an alternating voltage.

There are multitudes of alternating currents and voltages encountered in electrical and electronic equipment. However, a-c circuit analysis is easiest when the *sine wave* is chosen as the elementary waveform. Therefore, most of the subsequent discussion in this text concerns the sine-wave response of various a-c circuits.

The instantaneous value of the voltage shown in Figure 20-1 is constantly changing. The peak voltage occurs at the top-most point of the sine wave. Of course, there is an equal negative peak voltage induced in the loop 180° later. *Note that instantaneous values are denoted by small letters.*

The equation of the sinusoidal voltage is written

$$e = E \sin 2\pi ft \qquad (20.1)$$

or

$$e = E \sin \omega t \qquad (20.2)$$

where e denotes that instantaneous voltage value, E denotes the peak voltage, f denotes the frequency in Hertz, and t denotes elapsed time in seconds.

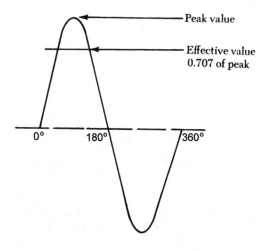

Figure 20–1
A sine-wave voltage.

Example 1 What is the instantaneous value of a 1 kHz sine wave whose peak voltage is 100 V at the time $(t) = 0.833$ msec?

$$e = E \sin 2\pi ft$$

$$e = 100 \sin 6.28 \times 1 \times 10^3 \times 0.833 \times 10^{-3}$$

$$e = 100 \sin 5.23 \text{ radians}$$

$$\sin(5.23 \text{ radians} \times 57.3°/\text{radians}) \approx \sin 300°$$

$$e = 100 \times -0.867$$

$$e = -86.7 \text{ V}$$

Problems 20–1

1. An alternating voltage has a peak value of 120 V and a frequency of 60 Hz. What is the instantaneous voltage at (a) $t = 0.12$ msec, (b) $t = 2.5$ msec, and (c) $t = 12$ msec?
2. An alternating voltage has a peak value of 120 V. What is its instantaneous value at (a) $\pi/4$ radians, (b) $\pi/6$ radians, (c) $\pi/3$ radians, (d) 25.5°, (d) 83°, and (f) 315°?
3. An alternating voltage has an instantaneous value of -150 V at 225°. What is its maximum value?
4. At what angle is the instantaneous value of a voltage equal to $1/\sqrt{2}$ of its peak voltage?
5. At what angle is the instantaneous value of a voltage equal to $2/\pi$ of its peak voltage?

 ## 20.2   Voltage Relations in a Sine Wave

We have learned the meaning of *instantaneous voltage*, and of *peak voltage*, in a sine wave. Since the positive-peak voltage is equal to the negative-peak voltage. The peak-to-peak voltage is equal to double the peak voltage.

Thus, we write

$$E_{pp} = 2E_p \qquad\qquad (20.3)$$

When we apply a sine-wave voltage to a resistor, a sine-wave current flows. The formula of this sine-wave current is written

$$i = I \sin \omega t \qquad\qquad (20.4)$$

Example 2 A 60-Hertz current has a peak value of 2 A. What is the instantaneous value of the current at $t = 0.02$ sec?

$$i = I \sin 2\pi ft$$

$$= 2 \sin 377 \times 0.02$$

$$= 2 \sin 7.54$$

$$7.54 \text{ radians} \times 57.3°/\text{radian} \approx 433°$$

$$i = 2 \sin 433° = 1.91 \text{ amp}$$

The power value in an a-c circuit is equal to the product of a-c voltage and a-c current. The number of watts is equal to the product of *effective* voltage and *effective* current. The effective values of a sine wave is equal to $1/\sqrt{2}$ or 0.707 of its peak value. Thus, if a sine-wave voltage has a peak value of 10 V, it has an effective value of 7.07 V; if a sine-wave current has a peak value of 10 A, it has an effective value of 7.07 A. The effective value is also called *the root-mean-square value*. Most VOMs and DVMs are calibrated to read effective voltage values of sine waveforms.

The power formula for a-c states

$$P = E_{eff}I_{eff} \qquad (20.5)$$

where P is the power in watts, E_{eff} is the effective voltage, and I_{eff} is the effective current.

Note very carefully that Formula 20.5 implies that the voltage and current waveforms are *in phase*; in other words, the phase difference between voltage and current waveforms is 0°.

To summarize:

$$E_{eff} = \frac{(E_{peak})^2}{2} \qquad (20.6)$$

$$E_{eff} = \frac{E_{peak}}{\sqrt{2}} = 0.707E_{peak} \qquad (20.7)$$

Problems 20–2

1. The peak value of a 2 kHz current is 200 mA. What is the instantaneous value of the current at time = 0.1 msec, $\theta = 160°$?
2. The instantaneous value of a 10 kHz current is 8.6 mA at $t = 30$ μsec.
   (a) What is the maximum value of the current?
   (b) What is the effective value of the current that would be measured by an ammeter?
   (c) What is the average value of the current?
3. The peak values of current and voltage through a resistor are found to be 250 μA and 3 kV, respectively.
   (a) What is the dissipated power in watts?
   (b) What is the value of the resistor in ohms?
4. An rms voltmeter indicates 162 V.
   (a) What is the peak value of the voltage?
   (b) What value would be measured on an oscilloscope (peak-to-peak)?
5. An ammeter measuring an a-c current indicates 112 μA. What is the peak value of the current?
6. The input signal current to a transistor amplifier has a value of 12 μA Rms. What is its peak-to-peak value?
7. A power outlet supplies 120 V rms; what is the peak-to-peak voltage?

Analysis of a-c series resistive circuits is basically the same as for d-c series resistive circuits. However, we will make progressive extensions of d-c concepts in the mathematical treatment of a-c circuitry as other components are added.

The power dissipated in a resistor is defined as the product of the rms voltage and the rms current, or as the square of the rms current multiplied by the resistance value, or as the square of the rms voltage divided by the resistance value.

In electronics work, we are sometimes concerned with the peak power that is developed. Peak power is formulated:

$$P_{peak} = I_{peak}E_{peak} \qquad (20.8)$$

We may connect resistors in any configuration across an a-c generator, and analyze the circuit in the same manner as if it were a d-c circuit; we merely state the voltage and current values in rms, peak, or peak-to-peak units, recognizing that power values in equivalent d-c watts are obtained only when we state the voltage and current in rms units.

 ## 20.3   Capacitors in A-C Circuits

A capacitor imposes an *opposition* to current flow that can be compared with resistance in a d-c circuit. However, since the current flow leads to the applied voltage by 90°, we recall this opposition to current flow *reactance*, or *capacitive reactance*. Reactance is measured in ohms, just as resistance is measured in ohms. However, the ohmic value of a capacitive reactance varies inversely to both the value of frequency and capacitance. The formula for capacitive reactance is:

$$X_C = \frac{1}{2\pi fC} \tag{20.9}$$

where X_C is the value of capacitive reactance in ohms, f is the frequency in Hertz, and C is the value of the capacitor in farads.

The current that flows in a capacitive circuit in which an a-c sine wave voltage is applied leads the voltage by 90°. The current is calculated by dividing the voltage by the capacitive reactance.

$$I = \frac{E}{X_C} \tag{20.10}$$

Example 3   What value of current flows in a circuit when a 100 V 1.59 kHz sine wave is applied to a 200 pF capacitor?

$$X_C = \frac{1}{2\pi fC} = \frac{0.159}{1.59 \times 10^3 \times 2 \times 10^{-10}} = 0.05 \times 10^7 = 500 \text{ k}\Omega$$

$$I = \frac{E}{X_C} = \frac{100}{500 \times 10^3} = 0.2 \text{ mA or } 200 \text{ } \mu\text{A}$$

Problems 20–3

1. A 115 V sine wave is applied across a 1 kΩ resistor. What value of current flows through the resistor? What amount of power is dissipated in the resistor?
2. Two 330 Ω resistors are connected in series across a 110 V a-c line; what is the total circuit current and the power across each resistor?
3. Two 47 kΩ resistors are connected in parallel across an alternating-voltage generator. The current in one resistor has a value of 100 μA. What current and power values are delivered by the generator?
4. What is the value of the peak current in Problem 3?
5. A 0.01 μF capacitor is connected across a 120 V, 60 Hz energy source; what is the value of the current flow in the circuit?
6. What are the instantaneous current and voltage values in the circuit in Problem 5 when θ is +30° and t is 0.1 sec?
7. The current through a 0.01 μF capacitor in a circuit is 100 μA.
   (a) What value of current would flow if the value of the capacitor was increased to 0.02 μF?
   (b) What value of current would flow if the value of the capacitor was decreased to 0.005 μF?

8. The voltage across a capacitor measured with an oscilloscope is 122 V peak-to-peak.
   (a) What is the rms voltage?
   (b) What is the current flow if the value of the capacitor is 100 nF and frequency 60 Hz?
   Note: nano farad (nF) is 10^{-9} F.

 ## 20.4 Power Relations in Capacitor Circuits

Power is defined as the product of voltage and current. When the voltage and current are not in phase, the product of instantaneous values depends on the phase angle between voltage and current. First, let us consider the power developed in a resistor when a sine-wave voltage is applied. The power is entirely positive, because the current is positive when the voltage is positive, and the current is negative when the voltage is negative. In turn, their product is always positive.

On the other hand, in a capacitor the current is not always positive when the voltage is positive, and the current is not always negative when the voltage is negative. Therefore, the power waveform is half positive and half negative. Therefore, the average power in a capacitor is equal to zero, or there is no power dissipated in the capacitor.

We call the $\cos \theta$ between the current and voltage in a reactance circuit the *power factor*. Its value happens to be zero for a capacitor. However, the power factor might have any value from 0 to 1 in other circuits.

 ## 20.5 Capacitors Connected in Series

When two capacitors are connected in series the source voltage is equal to the sum of the voltage drops across capacitors, in accordance with Kirchhoff's voltage law. The larger voltage drops across the smaller capacitor because it has the highest reactance. The formula for capacitors in series is:

$$X_C = \frac{C_1 \times C_2}{C_1 + C_2} \tag{20.11}$$

Example 4   What is the equivalent capacitance of a 0.01 μF capacitor and a 0.022 μF connected in series?

$$X_C = \frac{1 \times 10^{-8} \times 2.2 \times 10^{-8}}{1 \times 10^{-8} + 2.2 \times 10^{-8}} = 0.688 \times 10^{-8} \text{ F} = 0.00688 \ \mu\text{F}$$

On the other hand, Formula 20.12 will be found more convenient when several series-connected capacitors are to be reduced to an equivalent capacitor. For example, suppose that we have four series-connected capacitors; their equivalent capacitance is formulated

$$\frac{1}{C_T} = \frac{1}{C_1} + \frac{1}{C_2} + \frac{1}{C_3} + \frac{1}{C_4} \tag{20.12}$$

Example 5   What is the equivalent capacitance of four capacitors whose values are 0.01 μF, 0.02 μF, 0.25 μF, and 0.0333 μF connected in series?

$$\frac{1}{C_T} = \frac{1}{1 \times 10^{-8}} + \frac{1}{2 \times 10^{-8}} + \frac{1}{25 \times 10^{-8}} + \frac{1}{3.33 \times 10^{-8}}$$

$$\frac{1}{C_T} = 1 \times 10^8 + 0.5 \times 10^8 + 0.04 \times 10^8 + 0.3 \times 10^8 = 1.84 \times 10^8$$

$$C_T = \frac{1}{1.84 \times 10^8} = 0.0054 \ \mu F$$

Problems 20–4

1. A 0.15 μF capacitor and a 0.22 μF capacitor are connected in series. What is their equivalent capacitance?
2. A 2 μF capacitor is connected in series with a capacitor of unknown capacitance. If the total capacitance is found to be a value of 0.86 μF, what is the value of the unknown capacitor?
3. Determine the equivalent capacitance of three 15-pF capacitors connected in series.
4. What current would flow in two capacitors, with values of 15 pF and 60 pF, connected in series across a 100 V, 100 kHz source of energy?

 ## 20.6 Resistance and Capacitance Connected in Series

We know that when a sine-wave voltage is applied across a resistor, a sine-wave current flows. On the other hand, when a sine-wave voltage is applied across a capacitor, a cosine-wave current flows. In turn, we will expect that when a sine-wave voltage is applied across R and C connected in series as depicted in Figure 20-2, the current will lead the applied voltage by less than 90°, but by more than 0°. Current flow is opposed by both the reactance of the capacitor and the resistance of the resistor. The voltage drop across the resistor is in phase with current and will lead the voltage drop across the capacitor by 90°.

Since the amount of current that flows in the series RC circuit depends upon both resistance and reactance, we give the combination of resistance and reactance a name: *impedance*. Its symbol is Z. Impedance is measured in ohms. A visualization of impedance variation versus frequency for a series RC circuit is shown in Figure 20-2.

The value of impedance is found by applying the Pythagorean theorem:

$$Z = \sqrt{R^2 + X^2} \ \Omega \tag{20.13}$$

Example 6 Determine the impedance and phase angle of a series RC circuit when $R = 3 \ k\Omega$ and $X_C = 4 \ k\Omega$.

$$Z = \sqrt{(3 \times 10^3)^2 + (4 \times 10^3)^2} = \sqrt{25 \times 10^6} = 5 \ k\Omega$$

$$\tan \theta = \frac{X_C}{R} = \frac{4}{3} = 1.333$$

$$\theta = -53.2°$$

Note: $-53.2°$ indicates a lagging phase angle.

$$Z = 5 \ k\Omega \angle 53.2°$$

The current that flows in the circuit is calculated from Ohm's law:

$$I = \frac{E}{Z} \tag{20.14}$$

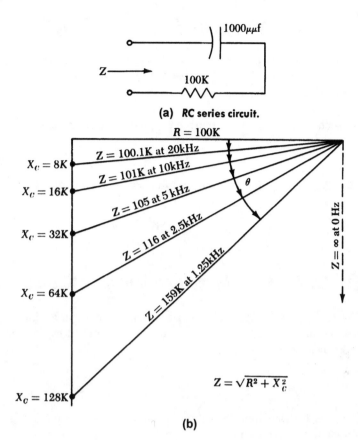

Figure 20–2
Voltage, current, resistance, reactance, and impedance relations in a series *RC* circuit. (a) *RC* series circuit.
(b) Impedance variation as the frequency is successively doubled.

Note that the vector diagram in Figure 20-2(a) shows the voltage and current relations in the circuit. If each vector is divided by the value of *I*, we obtain the impedance triangle. The impedance triangle shows the relations of resistance, reactance, and impedance. In this example, *I* = 1A. Both the vector diagram and impedance triangle show the *power-factor angle* θ. This is the angle between the resistance and impedance vectors.

Problems 20–5

1. A series circuit comprises a 200 Ω resistor and a 1 μF capacitor. What are the circuit impedance and phase angle when 300 V at 200 Hz is applied to the circuit?
2. What value of current flows in Problem 1, and what is the value of the power dissipated in the circuit in watts?
3. Calculate the circuit current and the voltage across a 2 μF capacitor and a 100 Ω resistor connected in series across a 20 V, 1 kHz generator.
4. A 12 V a-c, 1 kHz source is connected to a series circuit comprised of a 10 kΩ resistor and a 0.01 μF capacitor. What is the voltage across the resistor and the capacitor, and what is the total impedance?
5. What is the phase angle and the total current of the circuit in Problem 4?

Principles of Alternating Currents ▶ **161**

20.7 Principles of Inductive Reactance

An inductance imposes opposition to a-c current flow. This opposition, called *inductive reactance* is measured in ohms. If a *coil has an inductance of 1 H, a change in current flow of 1 amp per second causes a voltage drop of 1 V across the coil.* The current through an inductor lags the voltage by 90°.

When sine-wave voltage is applied to pure inductance, the formula for inductive reactance is

$$X_L = 2\pi fL = \omega L \ \Omega \tag{20.15}$$

Example 7 Determine the inductive reactance of a 100 mH inductor with an applied frequency of 159 kHz.

$$X_L = 6.28 \times 1.59 \times 10^5 \times 1 \times 10^{-1} \approx 10 \times 10^4 = 100 \text{ k}\Omega$$

 ## 20.8 Resistance and Inductance Connected in Series

Next, let us consider the voltage and current relations in a series *RL* circuit, such as the one depicted in Figure 20-3. The current flow is determined by the impedance of the circuit. The impedance value is formulated

$$Z = \sqrt{R^2 + X_L^2} \tag{20.16}$$

Example 8 Determine the value of the total impedance, phase angle, true power, and the total current of the circuit depicted in Figure 20-3.

$$X_L = 2\pi fL = 6.28 \times 1 \times 10^4 \times 0.159$$
$$= 1 \times 10^4 = 10 \text{ k}\Omega$$
$$Z = \sqrt{R^2 + X_L^2} = \sqrt{(1 \times 10^4)^2 + (1 \times 10^4)^2}$$
$$= \sqrt{2 \times 10^8} = 14.14 \text{ k}\Omega$$
$$\tan \theta = \frac{X}{R} = 1$$
$$\theta = 45°$$
$$I = \frac{E}{Z} = \frac{100 \text{ V}}{1.414 \times 10^4} = 7.07 \text{ mA}$$

In the example of Figure 20-3, the current value of 7.07 amp lags the applied voltage by 45°. The voltage drop across the inductor has a value of 70.7 V, and leads the applied voltage by 45°. The voltage drop across the resistor has a value of 70.7 V and lags the applied voltage by 45°. Thus, the voltage drop across the resistor lags the voltage drop across the inductor by 90°. Note that the power-factor angle θ is 45°.

Problems 20–6

1. A 10 V, 10 kHz source is connected across a 2.8 mH pure inductor. What is the current flow in the inductor?

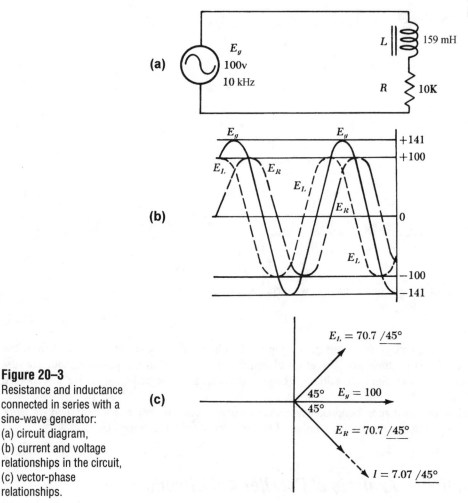

Figure 20–3
Resistance and inductance
connected in series with a
sine-wave generator:
(a) circuit diagram,
(b) current and voltage
relationships in the circuit,
(c) vector-phase
relationships.

2. A 25 V, 60 Hz source is connected across a 2.7 H inductor which has an internal resistance
    of 212 Ω. What is the inductor's impedance, the total circuit current, the circuit's phase
    angle, the circuit's true power, and the circuit's apparent power?
3. A 200 mV, 1 MHz source is connected across a 100 μH inductor in series with a 500 Ω
    resistor. What is the value of the total circuit impedance and current?
4. Suppose that the frequency in Problem 3 is reduced to 159 kHz. What is the new value of
    circuit impedance and current?
5. A 230 V, 60 Hz source is furnishing energy for a motor. If the current and phase angle
    are measured as 3 amp and 12°, respectively, what is the value of the motor's internal
    inductance and resistance?

▶ 20.9 Admittance and Susceptance

We know that conductance is the reciprocal of resistance. Thus $G = 1/R$. In the same way,
susceptance is the reciprocal of reactance. Thus, $B_L = 1/X_L$, or $B_C = 1/X_C$. In j notation, we

write for inductance

$$B_L = \frac{1}{jX_L} \qquad\qquad (20.17)$$

for capacitance

$$B_C = \frac{1}{-jX_C} \qquad\qquad (20.18)$$

In the same way, admittance is the reciprocal of impedance.

$$Y = \frac{1}{Z} \qquad\qquad (20.19)$$

In the case of inductive admittance, we write

$$Y = G - jB_L \qquad\qquad (20.20)$$

In the case of capacitive admittance, we write

$$Y = G + jB_C \qquad\qquad (20.21)$$

Note very carefully that the sign of j must be changed whenever we transfer j from the denominator to the numerator. Therefore, although inductive reactance is positive, inductive susceptance is negative. Similarly, although capacitive reactance is negative, capacitive susceptance is positive.

The use of admittance, conductance, and susceptance units is very expedient when we work with reactive circuits that have a number of components connected in parallel.

 ## 20.10 Analysis of Parallel A-C Circuits

Analysis of a-c resistive parallel circuits is similar to analysis of d-c resistive parallel circuits; the chief distinction is that the values of a-c voltage, current, and power can be expressed in various units. When a 120 V sine-wave voltage is applied to three parallel-connected resistors the applied voltage has an rms value of 120 V. In turn, its peak value is 169 V, approximately. The current flow through each resistor follows directly from Ohm's law.

The total capacitance of capacitors connected in parallel is the sum of their values.

$$C_{eq} = C_1 + C_2 + C_3 \qquad\qquad (20.22)$$

where C_{eq} denotes the equivalent capacitance of the three parallel-connected capacitors.

The current flow *through* each capacitor depends upon its reactance.

Example 9   Determine the equivalent capacitance of three capacitors, a 1 μF, a 2 μF, and a 5 μF connected in a parallel circuit.

$$C_T = 1 \ \mu\text{F} + 2 \ \mu\text{F} + 5 \ \mu\text{F} = 8 \ \mu\text{F}$$

It follows that the equivalent or total reactance of the circuit in Example 11 formulated

$$X_{Ctotal} = X_{C1} + X_{C2} + X_{C3} \qquad\qquad (20.23)$$

susceptance is

$$B_C = \frac{1}{X_C} = 2\pi fC \tag{20.24}$$

we may write

$$B_{eq} = B_1 + B_2 + B_3 \tag{20.25}$$

Current flow is evidently equal to

$$I_T = EB_{eq} \tag{20.26}$$

Problems 20–7

1. Two capacitors, 0.002 μF and 0.0015 μF, are connected in parallel. What is their total equivalent value?
2. Determine the total susceptance of two capacitors, 100 pF and 220 pF, connected in parallel at a frequency of 159 kHz.
3. An air-dielectric capacitor has 11 pF of capacitance for each pair of plates. What total number of plates is necessary for the capacitor to have a capacitance of 121 pF?
4. The total susceptance of four capacitors of equal capacitance connected in parallel is 1 mS at 15.9 kHz. What is the capacitance of each capacitor?

 ## 20.11   Resistance and Capacitance Connected in Parallel

Next, let us analyze the action of a circuit that comprises resistance and capacitance connected in parallel, as depicted in Figure 20-4(a). The same voltage is applied across the resistor and across the capacitor. The circuit is analyzed with the application of the following formulas.

$$Y = \sqrt{G^2 + B^2} \tag{20.27}$$

$$Z = \frac{1}{Y} \tag{20.28}$$

$$I_T = \sqrt{I_R^2 + I_C^2} \tag{20.29}$$

$$\tan \theta = \frac{B_C}{G} \text{ or } \frac{I_C}{I_R} \tag{20.30}$$

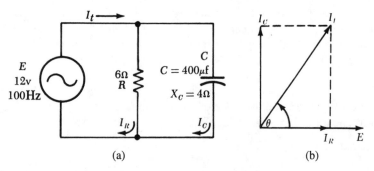

Figure 20–4   Parallel RC circuit: (a) Circuit diagram, (b) Vector diagram of I_C, I_R, and I_T.

Example 10 Find the equivalent impedance of a parallel *RC* circuit in which a supply voltage of 110 V results in values of $I_R = 30$ mA and $I_C = 40$ mA.

$$I_T = \sqrt{30^2 + 40^2} \text{ mA} = \sqrt{900 + 1600} \text{ mA} = 50 \text{ mA}$$

$$Z = \frac{E}{I_T} = \frac{110}{0.05} = 2,200 = 2.2 \text{ k}\Omega$$

The phase angle of the circuit in relation to admittance is 34.6°, and the phase angle of the circuit in relation to impedance is −34.6°. Unless stated otherwise, the term phase angle, by convention, refers to the impedance phase angle.

Example 11 Determine the impedance of a 100 Ω resistor connected in parallel with a capacitor with $X_C = 50$ Ω using conductance and admittance values.

$$G = \frac{1}{R} = \frac{1}{100} = 0.01 \text{ S} = 10 \text{ mS}$$

$$B_C = \frac{1}{X_C} = \frac{1}{50} = 0.02 \text{ S} = 2 \text{ mS}$$

$$Y = \sqrt{G^2 + B^2} = \sqrt{10^2 + 2^2} \text{ mS} = \sqrt{104} \text{ mS} \approx 10.2 \text{ mS}$$

Phase angle for the parallel circuit is

$$\tan \theta^{-1} = \frac{B_C}{G} = \frac{2 \text{ mS}}{10 \text{ mS}} = 0.2$$

$$\theta = +11.3°$$

The equivalent series circuit (ESC) is

$$Z = \frac{1}{Y} = \frac{1}{0.102}\angle 11.3° \cong 98\angle 11.3°$$

$$\tan \theta = \frac{X_{CS}}{R_s} =$$

$$X_C = Z \sin \theta = 89.3 \sin(-57.1°) = 89.3 \times 0.84 = -j75 \ \Omega$$

$$R = Z \cos \theta = 89.3 \cos(-57.1°) = 89.3 \times 0.54 = 48.5 \ \Omega$$

$$Z = 45.5 + j75 \ \Omega$$

Note: 89.5 Ω is the equivalent series impedance comprised of an equivalent series resistor and series capacitor. *Anytime we refer to impedance we are speaking of a series circuit.* Note that the ESC is valid for only one frequency. If the frequency of the input to the parallel circuit were to change, so would the ESC.

Exercises 20–1 Determine the value of the total impedance, phase angle, and equivalent series circuit for the following parallel circuits.

1. $R = 100$ kΩ and $X_C = 150$ kΩ
2. $R = 10$ kΩ, $C = 100$ pF, and $f = 100$ kHz
3. $G = 100$ μS and $B_C = 3.5$ μS
4. $G = 455$ μS and $B_C = 45.5$ μS

5. $G = 315\ \mu S$ and $\theta = 30°$
6. $G = 1$ mS and $B_C = 1$ mS
7. $R = 6.8$ kΩ and $X_C = 2.2$ kΩ

 ## 20.12   Resistance and Inductance Connected in Parallel

When resistance and pure inductance are connected in parallel the circuit action can be compared in many respects with a parallel configuration of resistance and capacitance. However, two basic distinctions must be kept in mind; the total current lags the applied voltage, and the equivalent impedance approaches the value of the shunt resistance as the frequency is increased indefinitely. The phase angle of the impedance is positive and the phase angle of the admittance is negative.

The value of Y is formulated

$$Y^2 = G^2 + B_L^2 \qquad\qquad (20.31)$$

Problems 20–8

1. A 470 Ω resistor and a 1.5 H coil are connected in parallel across a 110 V, 60 Hz source. Determine the equivalent circuit impedance, circuit phase angle, and the equivalent series circuit.
2. A parallel circuit composed of a 100 Ω resistor and a 2 mH coil is connected across a 100 mV, 15.9 kHz source. What is the total circuit current?

Determine the value of the impedance, circuit phase angle, and power dissipated in the following parallel circuits.

3. $R = 10$ kΩ, $X_L = 50$ kΩ, and $E_S = 110$ V
4. $G = 10$ mS, $R_L = 1$ mS, and $E_S = 200$ mV
5. $G = 10$ mS, $B_L = 147$ mS, and $E_S = 25$ V
6. $X_L = 47\ \Omega$, $R = 10$ kΩ, and $E_S = 56$ V
7. $G = 14.7$ mS, $B_L = 14.7$ mS, and $I = 10\ \mu A$

 ## 20.13   The j Operator Notation

We are now in position to consider calculations with the j operator notation as applied to reactive circuit analysis. A comprehensive discussion of the j operator was covered in Chapter 15. Since multiplication by the j operator rotates a vector counterclockwise by 90°, it follows that we may formulate the impedance of a series RL circuit as follows:

$$Z = R + jX_L \qquad\qquad (20.32)$$

Similarly, we may formulate the impedance of an RC circuit as follows:

$$Z = R - jX_C \qquad\qquad (20.33)$$

Thus, from the standpoint of j-operator notation, inductive reactance is positive, and capacitive reactance is negative. Formula 20.32 states that inductive reactance is 90° counterclockwise from resistance. Conversely, Formula 20.33 states that capacitive reactance is 90° clockwise from resistance. These relations are depicted in Figure 20-5(b) and (d).

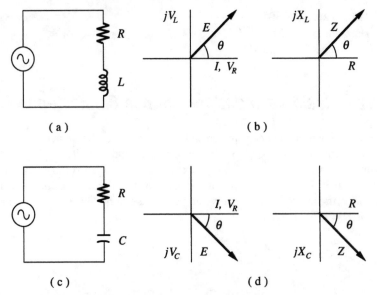

Figure 20–5 A series LR circuit: (a) Inductance and resistance connected in series, (b) Vector diagram in j notation, (c) Capacitance and resistance connected in series, and (d) Vector diagram in j notation.

Next, let us formulate the susceptance and phase angle of the parallel circuit depicted in Figure 20-6(a), using j notation.

$$Y = G - jB \text{ Siemen} \qquad (20.34)$$

$$Y = \sqrt{G^2 + B^2} \qquad (20.35)$$

$$\tan \theta = \frac{B}{G} \qquad (20.36)$$

The conductance and susceptance diagram of a parallel RC circuit follows the convention of a current vector diagram.

Exercises 20–2 Determine the value of the impedance, the admittance, and the phase angle of total current versus voltage of the following parallel circuits.

1. $R = 100\ \Omega, X = j150\ \Omega$
2. $R = 150\ \Omega, X = -j300\ \Omega$
3. $R = 17.3\ \text{k}\Omega, X = j10\ \text{k}\Omega$
4. $R = 3\ \text{k}\Omega, X = j4\ \text{k}\Omega$
5. $G = 330\ \mu\text{S}, B = j280\ \mu\text{S}$

Figure 20–6
Resistance and capacitance connected in parallel with a constant-current source:
(a) Circuit diagram,
(b) Admittance, conductance, and susceptance diagram.

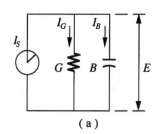

(a)

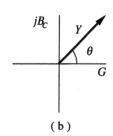

(b)

6. $G = 15.5 \text{ mS}, B = -j30 \ \mu\text{S}$
7. $G = 213 \ \mu\text{S}, B = j455 \ \mu\text{S}$
8. $G = 1.47 \ \mu\text{S}, B = j4.55 \ \mu\text{S}$
9. $G = 10 \ \mu\text{S}, B = -j31.5 \ \mu\text{S}$
10. $G = 455 \ \mu\text{S}, B = -j213 \ \mu\text{S}$

 ## 20.14 Analysis of Series-Parallel LCR Circuits

When an inductor and capacitor are connected in series with a resistor as shown in Figure 20-7, the impedance changes with frequency as the reactance of the inductor and capacitor change.

For the series circuit impedance

$$Z = \sqrt{R^2 + (X_L - X_C)^2} \tag{20.37}$$

for the phase angle

$$\tan \theta = \frac{X}{R} \tag{20.38}$$

Example 12 Determine the impedance, phase angle, and ESC for the LCR circuit shown in Figure 20-7.

$$X_L = 2\pi fL = 6.28 \times 10^4 \times 1 = 6.28 \times 10^4 = j62.8 \text{ k}\Omega$$

$$X_C = \frac{1}{2\pi fC} = \frac{1}{6.28 \times 10^4 \times 1 \times 10^{-9}} = -j15.9 \text{ k}\Omega$$

$$Z = \sqrt{(2 \times 10^4)^2 + (6.28 \times 10^4 - 1.59 \times 10^4)^2} = \sqrt{2^2 + 4.69^2} \times 10^4$$

$$Z = \sqrt{4 + 22} = \sqrt{26} = 5.1 \times 10^4$$

$$\tan \theta = \frac{X}{R} = \frac{4.69}{2} = 2.345$$

$$\theta = 66.9°$$

The ESC is comprised of a 20 kΩ resistor and an inductive reactance of 46.9 kΩ.

As the frequency approaches a point where $X_C = X_L$ the circuit becomes purely resistive and current is limited only by the resistor. The phase angle becomes 0°. A plot of current and phase angle is shown in Figure 20-7(c). This point is called resonance and denoted f_o and f_r.

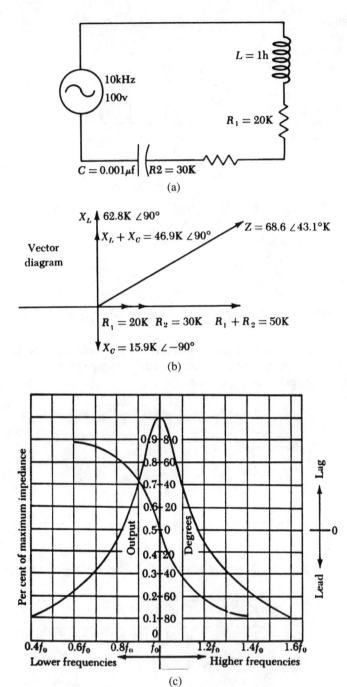

Figure 20–7
Series-parallel LCR circuit:
(a) circuit diagram, (b)
vector diagram for circuit
at 10 kHz, (c) current and
phase relationships.

The curve in Figure 20-7(c) represents either the circuit current relation or the circuit impedance characteristics. The curve could be plotted in decibels and at either side of the resonant have a point of −3 dB. These points are called cutoff frequency f_1 and f_2. The distance between the cutoff points and the shape of the bandwidth curve is determined by the circuit resistance or circuit Q. Circuit Q, like the Q of a coil is a figure of merit. A Q of ten or above is considered a high Q. The bandwidth of the circuit is the number of cycles between the cutoff points. At these points the circuit characteristics are:

At f_1

$$\theta = -45°$$

$$(X_C - X_L) = R$$

$$\text{Circuit power} = \frac{1}{2} \text{ power maximum}$$

$$V_R = (V_C - V_L) = 0.707 \, E$$

At f_2

$$\theta = +45°$$

$$(X_L - X_C) = R$$

$$\text{Circuit power} = \frac{1}{2} \text{ power maximum}$$

$$V_R = (V_L - V_C) = 0.707 \, E$$

At f_r

$$X_C = X_L$$

$$Z = \text{minimum} = R$$

$$\theta = 0°$$

$$I = \text{maximum}$$

$$f_r = \frac{1}{2\pi\sqrt{LC}} = \frac{0.159}{\sqrt{LC}} \qquad \textbf{(20.39)}$$

$$\text{Bandwidth} = f_2 - f_1 \qquad \textbf{(20.40)}$$

Note: $1/2\pi = 0.159$.

The resonant frequency of the circuit in Figure 20-7 is

$$f_r = \frac{1}{2\pi\sqrt{LC}} = \frac{0.159}{\sqrt{1 \times 10^{-9}}} = \frac{0.159}{\sqrt{.1 \times 10^{-8}}}$$

$$= \frac{0.159}{.316 \times 10^{-4}} \cong 5 \text{ k}\Omega$$

Exercises 20–3   Solve the following series LCR circuit for f_r, Z, Q, f_1, f_2, and bandwidth.

1. $L = 15.9$ mH, $C = 6.28 \, \mu\text{F}$, and $R = 200 \, \Omega$
2. $L = 100$ mH, $C = 0.001 \, \mu\text{F}$, and $R = 1000 \, \Omega$
3. $L = 10$ mH, $C = 100$ nF, and $R = 100 \, \Omega$

▶ *20.15 Analysis of Series-Parallel Circuits*

An RC series-parallel circuit is depicted in Figure 20-8(a). To determine the circuit impedance, phase angle, circuit current, and equivalent series circuit (ESC) we proceed as follows.

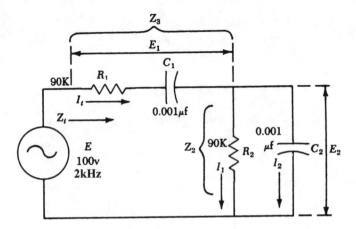

Figure 20–8
Series-parallel circuit
diagram.

1. Convert the parallel part of the circuit to its ESC.
2. Add the components of the series part to the ESC.
3. Solve the ESC for Z, I, power dissipation, etc.

Example 13 Find angle θ, ESC, and I_t for the circuit in Figure 20-8.

(a) Calculate the reactance of the series capacitor and of the parallel capacitor.

$$X_{C1} = X_{C2} = \frac{1}{6.28 \times 2 \times 10^3 \times 1 \times 10^{-9}} = 79.5 \text{ k}\Omega$$

(b) Formulate the impedance of R_1 and C_1 connected in series.

$$Z_3 = R - jX_C$$
$$Z_3 = 90K - j79.5 \text{ k}\Omega$$

(c) Formulate the impedance of R_2 and C_2 connected in parallel into an equivalent series impedance.

$$Z_2 = \frac{-jRX_C}{R - jX}$$
$$Z_2 = \frac{(90 \text{ K})(79.5 \text{ K}\angle-90°)}{90 \text{ K} - j79.5 \text{ K}} \Omega$$

(d) Calculate the impedance as seen from the generator.

$$Z_T + Z_T$$

We must change the parallel impedance comprised of C_2 and R_2 to a series equivalent circuit. Then we can add Z_1 and Z_2.

$$Z_1 = R_1 + C_1 = 90 \text{ k}\Omega - j79.5 \text{ k}\Omega$$
$$Z_2 = R_2 \parallel C_2 = \frac{R_2 \times (-jX_{C2})}{R_2 + (-jX_{C2})}$$

$$= \frac{90 \times 10^3 \times -j79.5 \times 10^3}{90 \times 10^3 - j79.5 \times 10^3}$$

$$= \frac{(90 \times -j79.5) \times 10^6}{(90 - j79.5) \times 10^3}$$

Cancel powers of ten and multiply both the numerator and denominator with a conjugate to clear the denominator of the j term.

$$= \frac{(90 \times -j79.5)(90 + j79.5) \times 10^3}{(90 - j79.5)(90 + j79.5)}$$

$$= \frac{(-j643950 - j^2 56882) \times 10^3}{14420} = 39.4\,k - j44.6\,k$$

$$Z_1 + Z_2 = 90\,k - j79.5\,k + 39.4\,k - j44.6\,k$$

$$Z_T = 129.5\,k - j124.1\,k$$

or in the polar form

$$Z_T = \frac{179\,k}{-44°}\,\Omega$$

$$I_T = \frac{E}{Z} = \frac{100}{179\,k\angle -44°} \approx 599\,\mu A$$

Problems 20–9   Make a drawing of each of the circuits pertaining to these problems before attempting to find a solution.

1. Two impedances, Z_1 and Z_2, are connected in series; $Z_1 = 141\angle -45°$ and $Z_2 = 200\angle +36.8°$. What is the total impedance of the network?
2. Two series networks, Y_1 and Y_2, are connected in parallel; $Y_1 = 100\angle 36°\,\mu S$ and $Y_2 = 45.5\angle -50°\,\mu S$. Determine Y_T, Z_T, θ, and the current through each parallel branch of the circuit when 220 V are applied across the network.
3. Determine the equivalent impedance for two impedances, Z_1 and Z_2, connected in parallel; $Z_1 = (200 + j5\,k)\,\Omega$ and $Z_2 = (-j4\,k)\,\Omega$.
4. Two circuits (Z and Y) are connected in series; $Z = (1\,k + j1\,k)\,\Omega$ and $Y = (100 + j200)\,\mu S$. What is their total impedance and phase angle (θ)?
5. Determine the equivalent impedance, the phase angle θ, and the total current of the circuit in Figure 20-8, when $f = 100\,kHz$, $L = 200\,\mu H$, $C = 200\,\mu F$, and $R = 500\,\Omega$.
6. Determine the value of the total impedance, phase angle, and power dissipated in the circuit of Figure 20-8, when $f = 159\,kHz$, $L = 6.28\,mH$, $C = 100\,pF$, and $R = 1000\,\Omega$.

In the preceding discussion of parallel-circuit analysis, we used the product-and-sum formula to find the total impedance of the shunt-connected components. In the following discussion of series-parallel circuit analysis, we will again encounter both vector products and vector quotients.

▶ 20.16 Analysis of Series-Parallel LCR Circuits

There are various configurations of series-parallel LCR circuits. The most typical type is shown in Figure 20-9. Observe that if R were equal to zero, the circuit would reduce to a configuration with only a capacitor and an inductor in parallel. However, the presence of R changes the circuit action

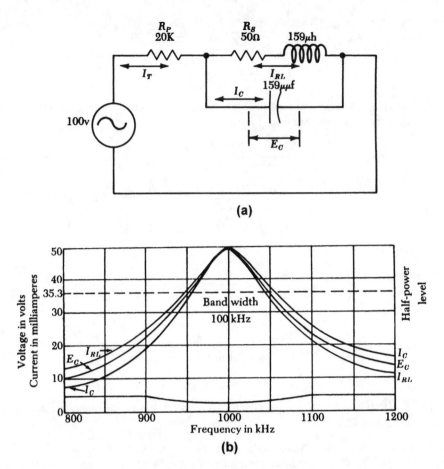

Figure 20–9 Series-parallel relationships: (a) circuit diagram, (b) current versus frequency diagram.

considerably. The value of R affects the minimum value of I_T, the bandwidth, and the resonant frequency, as will be explained. Variation in I_T for three values of R is shown in Figure 20-9(b). The reader will perceive that I_T lags at frequencies below resonance, and leads at frequencies above resonance. This is the opposite action from that of a series resonant circuit.

As with a series LCR circuit, we will find that the parallel resonant frequency can be defined in three different ways: 1) when I_T = maximum, 2) when Z = minimum, and 3) $\theta = 0°$. These three points occur at different frequencies. However they approach each other as resistance of the coil decreases, resulting in a simplified formula. This formula is correct for most calculations when the Q of the circuit is greater than 10.

$$f_r = \frac{1}{2\pi LC} \tag{20.41}$$

The value of Q is the current in the reactance divided by the line current when Q is calculated at f_r.

$$Q = \frac{I_X}{I_{Line}} \text{ or } \frac{G_p}{B} \tag{20.42}$$

At f_r, I_{LR} is Q times I_T; this is the current magnification of the parallel-resonant circuit. If R is small, we may formulate the circuit impedance at resonance as follows:

$$Z = \frac{L}{RC} \qquad \text{(20.43)}$$

or

$$Z = Q^2 R_p \qquad \text{(20.44)}$$

We must remember that Formulas 20.43 and 20.44 are good approximations only when R_s is comparatively small and the equivalent value of R_p is large.

The bandwidth of the resonant circuit is formulated

$$BW = f_2 - f_1 \qquad \text{(20.45)}$$

or

$$BW = \frac{f_r}{Q} \qquad \text{(20.46)}$$

The circuit can be converted to an equivalent parallel circuit of LCR by multiplying the inductors of resistance by $Q \times R_s$.

Example 14 Determine the resonant frequency, circuit Q, and the bandwidth of the circuit in Figure 20-9.

$$f_r = \frac{1}{2\pi\sqrt{LC}} = \frac{0.159}{\sqrt{159 \times 10^{-6} \times 159 \times 10^{-12}}}$$

$$= \frac{0.159}{\sqrt{159 \times 159 \times 10^{-18}}} = \frac{0.159}{159 \times 10^{-9}} = \frac{159 \times 10^6}{159} = 1\ \text{MHz}$$

$$X_C = X_L = 2\pi fL = 6.28 \times 1 \times 10^6 \times 1.59 \times 10^{-4} = 1\ \text{k}\Omega$$

$$Q = \frac{X_L}{R_s} = \frac{1000}{50} = 20$$

Transfer 50 Ω series to parallel

$$R_p = Q^2 R_s = 20^2 \times 50 = 20\ \text{k}\Omega$$

at f_r

$$Z = R_p = 20\ \text{k}\Omega$$

or

$$Z = \frac{L}{R_s C} = \frac{159 \times 10^{-6}}{50 \times 159 \times 10^{-12}} = 20\ \text{k}\Omega\angle 0°$$

or

$$Z = R_p$$

$$\text{Bandwidth} = \frac{f_r}{Q} = \frac{1000\ \text{kHz}}{20} = 50\ \text{kHz}$$

$$f_1 = f_r - \frac{BW}{2} = 1000\ \text{kHz} - \frac{50\ \text{kHz}}{2} = 1000\ \text{kHz} - 25\ \text{kHz} = 975\ \text{kHz}$$

$$f_2 = f_r + \frac{BW}{2} = 1000\ \text{kHz} + \frac{50\ \text{kHz}}{2} = 1000\ \text{kHz} + 25\ \text{kHz} = 1025\ \text{kHz}$$

Let us observe the effects on bandwidth of the circuit in Figure 20-9 for two different values of R_p. If R_p = 20 kΩ, the bandwidth is approximately 50 kHz. On the other hand, if R_p = 5 kΩ, and the circuit Q = 5, then the bandwidth is approximately 200 kHz.

$$f_r = \frac{1}{2\pi\sqrt{LC}} = \frac{0.159}{\sqrt{LC}}$$

Problems 20–10

1. What value of capacitance will produce resonance at 15.9 kHz with a 22 mH inductor in a parallel circuit?
2. What value of series resistance will produce a bandwidth of 4 kHz in the circuit of Problem 3?
3. An inductor of 159 mH with a Q of 80 is connected in series with a 1 nF capacitor.
   (a) What is the circuit resonance frequency?
   (b) What is the impedance of the circuit at resonance?
   (c) What is the current bandwidth?
4. A 10 mH inductor with a Q of 80 is connected in parallel with a 365 pf capacitor.
   (a) What is the resonance frequency of the circuit?
   (b) What is the current through the inductor if the generator current is 100 mA at resonance?
5. A capacitance of 1000 pF is connected across an inductor of 10 mH. What values of parallel resistance will produce a bandwidth of 8 kHz?

Summary

1. Series LCR circuits at resonance have the characteristics of: Minimum Z, maximum I, $\theta = 0°$.
2. Parallel LCR circuits at resonance have the characteristics of: Maximum Z, minimum I, $\theta = 0°$.
3. At the cutoff frequencies of LCR circuit, the characteristics are: $\theta = \pm 0°$, $\frac{1}{2}$ power, and -3 dB.

Chapter 21

Boolean Algebra

 ## 21.1   Introduction

Boolean algebra was introduced by George Boole in 1854. The rules, laws, and theorems in Boole's work were known as "logical algebra." In 1938, Claude Shannon of MIT applied Boole's work to the operation of switches and relay circuits. This was the foundation of later applications of *Boolean Algebra* to a systemic method of expressing and analyzing the operation of computer logic circuits in a mathematical manner.

In Chapter 5 we stated that in computer operation, a plus voltage (high level) would be considered as a "1" and no voltage (low level) would be considered as a "0." A high voltage is also considered the TRUE state and a low voltage is considered a FALSE state. We will continue these notations in this chapter.

A variable is represented as a capital letter such as A, B, C, or X. The complement or inverse of the variable can be indicated by a bar over the variable, or a prime symbol. We will use the bar notation in this text.

$$A = 1 \text{ then } \bar{A} = 0$$

or

$$A = 1 \text{ then } \bar{A} = 0 \tag{21.1}$$

Conversely if:

$$A = 0 \text{ then } \bar{A} = 1$$

or

$$\bar{A} = 0 \text{ then } A = 1 \tag{21.2}$$

The logical AND function is represented by a · between the variables $(A \cdot B)$ or by writing adjacent letters (AB). The latter notation is most often used. The logical OR function is represented by the + sign between the variables $(A + B)$.

 ## 21.2   Boolean Addition and Multiplication

Addition in Boolean algebra of two or more variables represented by 1 or zero is as follows:

Table 21–1
Boolean Addition

$0 + 0 = 0$
$0 + 1 = 1$
$1 + 0 = 1$
$1 + 1 = 1$

In the application of Boolean algebra to computer logic circuits addition is the same as the OR function. However, it is important to note the difference between the Boolean algebra addition $(1 + 1 = 1)$ and the binary addition $(1 + 1 = 10)$.

Multiplication in Boolean algebra applies the same rules as binary multiplication.

Table 21–2
Boolean
Multiplication

$$0 \cdot 0 = 0$$
$$0 \cdot 1 = 0$$
$$1 \cdot 0 = 0$$
$$1 \cdot 1 = 1$$

 21.3 Logic Expressions and Symbology

There are six basic computer logic circuits. Five of these are used as logic expressions in Boolean algebra.

The *driver* is a voltage or current amplifier, it does not change the level or property of the input.

The *inverter* or *NOT* circuit inverts or complements the input. If the input is A the output is NOT A and conversely, if the input is NOT A the output is A. This circuit is sometimes called a NOT symbol. The symbol and circuit action for an inverter is depicted in Figure 21-1.

Figure 21–1
The inverter symbol is a
small bead and may be
used with the driver
symbol.

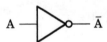

A ———▷∘— $\bar{A}$

The *OR* circuit or gate, depicted in Figure 21-2, can be expressed mathematically in equation form as: $A + B = X$. Where X is the result of the circuit action. The circuit symbol and circuit action are depicted below in Figure 21-2 and Table 21-3.

Table 21–3   A Summary of a Three Input OR and a
Three Input NOR Gate.

				OR CIRCUIT	NOR CIRCUIT
A	B	C	=	X	X
0	0	0	=	0	1
0	0	1	=	1	0
0	1	0	=	1	0
0	1	1	=	1	0
1	0	0	=	1	0
1	0	1	=	1	0
1	1	0	=	1	0
1	1	1	=	1	0

Figure 21-2
The symbols for a three input OR circuit.

When the OR function is inverted or complemented the result is that a "1" output is changed to a "0" or a "0" output is changed to a "1." In other words, a *high* output is changed to a *low* output and a *low* output is changed to a *high*. Figure 21-4 depicts the symbol for an NOR circuit. A summary of a three input NOR circuit function is shown in Table 21-3.

Figure 21-3
The circuit symbol for an NOR circuit.

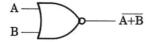

Figure 21-4
Circuit symbol for a three input AND gate.

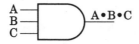

The AND gate requires that all inputs be high to produce an output. The operation of a three input AND circuit is summarized in Figure 21-4. The input variables are ABC. The output variable is in each case X. Table 21-4 summarizes the results of a three input AND gate.

Table 21-4   Table of Three Input AND and Three Input NAND Circuits.

		AND CIRCUIT			NAND CIRCUIT
A ·	B ·	C	=	X	$\overline{X}$
0	0	0	=	0	1
0	0	1	=	0	1
0	1	0	=	0	1
0	1	1	=	0	1
1	0	0	=	0	1
1	0	1	=	0	1
1	1	0	=	0	1
1	1	1	=	1	0

Problems 21-1

1. The inputs to an AND gate are: $A = 0, B = 1, C = 1$. What is the output?
2. An inverter is placed at the output of the AND gate in Problem 1; what is the output of the circuit?
3. The inputs to an OR circuit are $A = 1, B = 0, C = 0$; what is the output of the gate?
4. The inputs to an AND gate are $\bar{A}, B$, and $\bar{C}$; what is the output of the circuit?
5. Where would you place an inverter to change the level of the output of the gates in Problem 4?

▶ 21.4 Rules and Laws of Boolean Algebra

In areas of mathematics there are explicit laws and rules that must be properly followed for correct results. The same is true for solutions of Boolean algebra problems. Three of these laws are the same as in mathematics. These are the *Commutative laws*, the *Associative laws*, and the *Distributive laws*. In addition to these three fundamental laws there are self-evident truths which are called *Postulates*. The rules will be committed to memory as you work through the problems in this chapter.

Postulates   As we noted earlier, Postulates are truths that seem so self-evident that you might wonder why they are presented here in Table 21-5.

Table 21–5   Postulate properties

1a. $A = 1$ (if $\bar{A} = 0$)	1b. $A = 0$ (if $\bar{A} = 1$)
2a. $0 \cdot 0 = 0$	2b. $0 + 0 = 0$
3a. $1 \cdot 1 = 1$	3b. $1 + 1 = 1$
4a. $1 \cdot 0 = 0$	4b. $1 + 0 = 1$
5a. $1 = 0$	5b. $0 = 1$

Algebraic Laws   The following laws are properties of algebra that apply to Boolean algebra. The fundamental difference is that in Boolean algebra a variable may have the values of only 0 or 1.

Commutative Law   The commutative law for addition of two variables is written:

$$A + B = B + A \tag{21.3}$$

This states that the sequence of the variables applied to an OR gate make no difference in the output.

The commutative law for multiplication may be written:

$$AB = BA \tag{21.4}$$

This illustrates that the sequence of the variables to be ANDed makes no difference in the output.

Associative Law   The associative law for addition for three variables is:

$$A + (B + C) = (A + B) + C \tag{21.5}$$

This law states that the resulting output of an OR gate is the same regardless of the grouping of the variables.

The associative law is also applicable to AND gates. This law states that the resulting AND gate is the same regardless of the sequence of grouping of variables.

$$ABC = X$$
$$BAC = X \qquad \text{(21.6)}$$
$$CAB = X$$

Distributive Law   The distributive law is written:

$$A(B + C) = AB + AC \qquad \text{(21.7)}$$

The law states that ORing several variables and ANDing the result is the same as ANDing a single variable with each of the variables and ORing the products.

Table 21–6   A Summary Commutative Laws

6a. $AB = BA$	6b. $A + B = B + A$
7a. $A(BC) = (AB)C$	7b. $A + (B + C) = (A + B) + C$
8a. $A(B + C) = AB + AC$	8b. $A + BC = (A + B)(A + C)$

21.5   Theorems for Boolean Algebra

There are a number of theorems that can be applied to computer AND and OR logic gates to simplify the circuitry and reduce the number of gates. These theorems are listed in Table 21-7.

Table 21–7   Theorems for Boolean Algebra

1a. $A \cdot 0 = 0$	1b. $A + 0 = A$
2a. $A \cdot 1 = A$	2b. $A + 1 = 1$
3a. $A \cdot A = A$	3b. $A + A = A$
4a. $A \cdot \overline{A} = 0$	4b. $A + \overline{A} = 1$
5a. $\overline{\overline{A}} = A$	5b. $A = \overline{\overline{A}}$

Summary of Theorems

Rule 1a states that if one input of an AND gate is held at 0, the output will always be 0 or false.

Rule 1b states that if one side of an OR gate is held to the zero state, the output of the gate will always be A.

Rule 2a states that if one input of an AND gate is held at 1 or high, the output will be dependent only on the level of the other input (or inputs).

Rule 2b states that if one input of an OR gate is held at 1, the output of the gate will always be true (A) or 1 regardless of the other inputs.

Rule 3a states that if all inputs of an AND gate are true, the output will be true.

Rule 3b states that if both inputs of an OR gate are true, the output of the gate will always be true.

Rule 4a states that if the inputs to an AND gate are complementary (one high-one low), the output will always be low.

Rule 4b states that if the input to an OR gate are complementary, the output of the gate will always be high.

Rule 5a states that if a logic function is inverted twice, the result of the double negation will be the original level.

Rule 5b is a restatement of Rule 5a.

Let us use the Boolean theorems to simplify several logic expressions, express the results in logic diagrams, and finally develop a truth table for each.

Example 1 $A + AB = A$

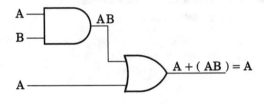

Truth Table

A	B	AB	A + AB
0	0	0	0
0	1	0	1
1	0	0	1
1	1	1	1

Figure 21–5
Logic diagram and truth table for Example 1.

Example 2 $A + ABC = A$

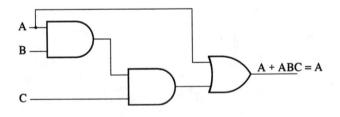

Truth Table

A	B	C	ABC	A + ABC
0	0	0	0	0
0	0	1	0	0
0	1	0	0	0
0	1	1	0	0
1	0	0	0	1
1	0	1	0	1
1	1	0	0	1
1	1	1	1	1

Figure 21–6
Logic diagram and truth table for Example 2.

Exercises 21–1 Make the truth table and draw the logical table for the following, simplifying where possible.

1.	$A + B$	**2.**	$A\bar{B}$
3.	ABC	**4.**	$A(B + C)$
5.	$AB + C$	**6.**	$AB + AC$
7.	$(A + B)(A + C)$	**8.**	$A + ABC$
9.	$A(\bar{B} + C)$	**10.**	$A\bar{B} + C\bar{D}$
11.	$AB + C$	**12.**	$(AB) + (A + B)$
13.	$ABC + AC + C$	**14.**	AB
15.	$A + \bar{B}$	**16.**	$ABC + 1$
17.	$AB + A\bar{C}$	**18.**	$AB + CD$
19.	$A + CD$	**20.**	$AB + CD + BC$

▶ **21.6 Demorgan's Theorems**

Demorgan proposed an extension of Boole's work in the form of two theorems that can be used to simplify Boolean equations.
    These are:

$$\bar{A}\bar{B} = \bar{A} + \bar{B} \tag{21.8}$$

$$A \mp B = \bar{A}\bar{B} \tag{21.9}$$

The first theorem can be stated, "The complement of a product of two or more variables is equal to the sum of the complement of the variables." This theorem is illustrated in Figure 21-7 with the accompanying truth table.

Figure 21–7
Logic diagram and truth table illustrating Demorgan's first theorem that $\bar{A}\bar{B} = \bar{A} + \bar{B}$.

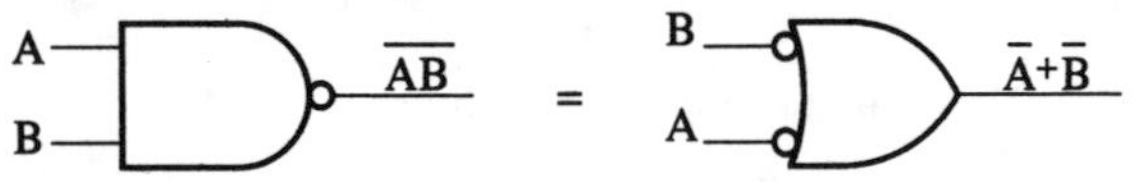

			Truth Table			
A	B	$\bar{A}$	$\bar{B}$	AB	$\bar{A}\bar{B}$	$\bar{A} + \bar{B}$
0	0	1	1	0	1	1
0	1	1	0	0	1	1
1	0	0	1	0	1	1
1	1	0	0	1	0	0

    The second of Demorgan's theorems states that the complement of the sum of two or more variables is equal to the product of the complement of each variable. This theorem is demonstrated in Figure 21-8 with an accompanying truth table.

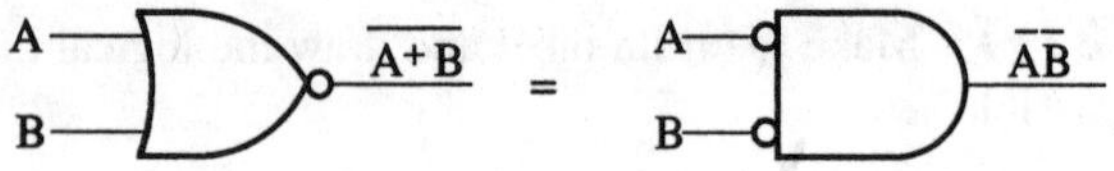

A ———|\\
      |)o— $\overline{A+B}$ A —o|\\
B ———|/ |)— $\overline{\overline{A}\,\overline{B}}$
                          =   B —o|/

Figure 21–8
Circuit diagram showing the relationship between a NOR circuit and the use of inverters with an AND circuit. Both circuits perform the NOR function.

				Truth Table		
A	B	$\overline{A}$	$\overline{B}$	$A + B$	$\overline{A + B}$	$\overline{A}\,\overline{B}$
0	0	1	1	0	1	1
0	1	1	0	1	0	0
1	0	0	1	1	0	0
1	1	0	0	1	0	0

Example 3 demonstrates the application of these two theorems.

Example 3 Simplify the term $\overline{A + B} + \overline{C + D}$.

1. Replace the $+$ sign with the $\cdot$ sign $\overline{(A + B) \cdot (C + D)}$.
2. Complement each of the two terms $\overline{(A + B)} \cdot \overline{(C + D)}$. (Demorgan's theorem)
3. Apply Rule 5a to remove the two bars $= (A + B)(C + D)$.

Example 4 Apply Demorgan's theorem to simplify the following expression.

$$\overline{A} + \overline{B} + \overline{C}$$

Step 1. Complement the entire expression.

$$\overline{\overline{A} + \overline{B} + \overline{C}}$$

Step 2. Complement each term.

$$\overline{\overline{\overline{A}} + \overline{\overline{B}} + \overline{\overline{C}}}$$

Step 3. Change the signs of each term.

$$\overline{\overline{\overline{A}} \cdot \overline{\overline{B}} \cdot \overline{\overline{C}}}$$

Step 4. Remove the double negation of each term.

$$\overline{A}\overline{B}\overline{C}$$

Exercises 21–2 Apply the Boolean laws and Demorgan's theorem to reduce the following expressions and draw the gate diagrams for each.

1. $\overline{A} + AC$
2. $AB + A\overline{C}$
3. $AB + A\overline{B}$
4. $AB + AC + \overline{A}D$
5. $(AB)(C + D)$
6. $(AB)(AC)$
7. $ABC + ABC + AC$
8. $(A + B + C)(ABC)$
9. $ABC(A + B + C)$
10. $(ABC)(ABC) + (A + B + C)$
11. $(ABC) + (\overline{A}B\overline{C})$
12. $ABC + \overline{A}BC + A\overline{B}\overline{C}$

21.7 KARNAUGH Maps

The first attempt of writing a digital equation may not produce the most efficient effort in terms of circuitry. KARNAUGH mapping is a rather simple method of reducing an equation to the lowest possible circuit combination. The arrangement is comprised of a tabulator grid in which the terms of the equation are entered.

The KARNAUGH map is comprised of a group of squares within a grid. The grid has 4 squares for a two variable equation, 8 squares for a three variable equation, and sixteen squares for a four variable equation. Figure 21-9 depicts the layout of a 2 variable AND equation. We recognize that the truth table in Figure 21-9(a) represents all of the conditions of the equation AB. The squares in Figure 21-9(b) represents all of the conditions of the AND equation.

0 01 1 1 0 0 1 1 0

Figure 21–9
Representation of a two
term Boolean equation:
(a) truth table,
(b) KARNAUGH map.

(a)

(b)

Figure 21-10 depicts the layout of a three term and a four term Boolean equation. The maps are arranged so that there is only a single variable change in any adjacent square. This layout is important and must be followed for correct results. Note that the three variable map has eight squares and the four variable map has 16 squares. Maps for greater numbers of variables are possible; however, those are usually developed on a computer because of their complexity.

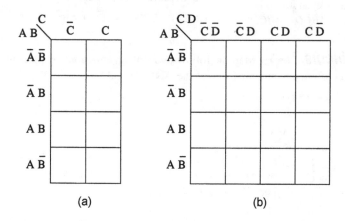

Figure 21–10
The layout of a three term
and a four term
KARNAUGH map.

(a)

(b)

 21.8 Entering Boolean Equations in a KARNAUGH Map

To enter the variables of a Boolean expression into a KARNAUGH map it must be in the *sum-of-products*. For example, the equation $ABC + ABC + ABC$ is entered in the map in Figure 21-11. There is a 1 square in the map that satisfies each AND portion of the equation.

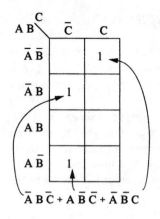

Figure 21–11
The KARNAUGH map for the equation
$\bar{A}B\bar{C} + A\bar{B}\bar{C} + \bar{A}\bar{B}C$.

$$\bar{A}\,B\,\bar{C} + A\,\bar{B}\,\bar{C} + \bar{A}\,\bar{B}\,C$$

Figure 21-12 represents the KARNAUGH map for the four variable Boolean term $ABCD + \bar{A}\bar{B}CD + \bar{A}\bar{B}\bar{C}D + \bar{A}\bar{B}\bar{C}\bar{D}$. Again, there is a 1 placed in each square that satisfies each AND portion of the equation.

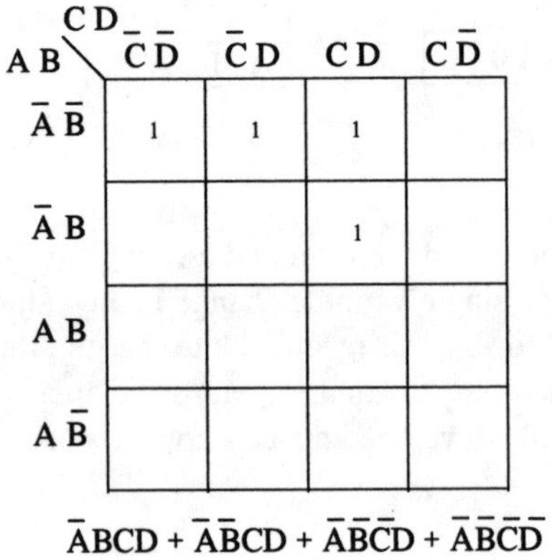

CD \ AB	$\bar{C}\bar{D}$	$\bar{C}D$	CD	$C\bar{D}$
$\bar{A}\,\bar{B}$	1	1	1	
$\bar{A}\,B$			1	
$A\,B$				
$A\,\bar{B}$				

Figure 21–12
The mapping of a four term Boolean expression for the equation $\bar{A}BCD + \bar{A}\bar{B}CD + \bar{A}\bar{B}\bar{C}D + \bar{A}\bar{B}\bar{C}\bar{D}$.

$$\bar{A}BCD + \bar{A}\bar{B}CD + \bar{A}\bar{B}\bar{C}D + \bar{A}\bar{B}\bar{C}\bar{D}$$

Problems 21–1 Map the following Boolean equations by placing a 1 in each of the squares that represent one and only one of the AND terms.

1. $AB + \bar{A}B$
2. $AB + A\bar{B}$
3. $A\bar{B}C + A\bar{B}C$
4. $AB\bar{C} + A\bar{B}C + \bar{A}\bar{B}C$
5. $AB\bar{C} + A\bar{B}C + \bar{A}BC + A\bar{B}\bar{C}$
6. $A\bar{B}\bar{C}\bar{D} + \bar{A}BCD + AB\bar{C}\bar{D}$
7. $\bar{A}\bar{B}\bar{C}\bar{D} + ABCD + A\bar{B}C\bar{D} + AB\bar{C}D$
8. $A\bar{B}C\bar{D} + \bar{A}B\bar{C}D + \bar{A}\bar{B}\bar{C}\bar{D} + ABCD$
9. $\bar{A}BC\bar{D} + A\bar{B}CD + \bar{A}B\bar{C}D + \bar{A}B\bar{C}\bar{D}$

▶ 21.9 Simplification of Boolean Equations

You have learned to locate the AND terms of Boolean *product-sum* equations on a KARNAUGH map. Now we are ready to use the map to simplify the equations. To accomplish this the 1s located in the squares are grouped. The groups include every one in a vertical or horizontal square (no diagonal groups). Each group includes the largest possible number of squares. Groups include 1s that are at the top and bottom of the vertical columns and the 1s that are at the left and right of the horizontal lines. This is because the map can be thought of as a vertical and a horizontal cylinder. To visualize this you might think of the map as a label that is removed from a can. To better visualize the grouping concept let us examine several examples. First, Figure 21-13 forms a single vertical group. We will develop this map later from a rather complex Boolean equation.

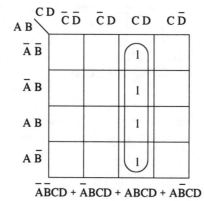

Figure 21–13
Only a single vertical group is formed from the equation $\bar{A}\bar{B}CD +$ $\bar{A}BCD + ABCD + A\bar{B}CD$.

$\bar{A}\bar{B}CD + \bar{A}BCD + ABCD + A\bar{B}CD$

Next let us examine the map shown in Figure 21-14. We note that there are three adjacent groups: three in the top horizontal row, three in the third horizontal row, and eight in the vertical rows. Remember, to be included in a group, the 1s must be adjacent in the 2 and 3 column.

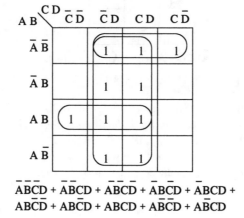

Figure 21–14
An example of a map with three groups for the equation
$\bar{A}\bar{B}\bar{C}D + \bar{A}\bar{B}CD +$
$\bar{A}BC\bar{D} + \bar{A}B\bar{C}D +$
$\bar{A}BCD + AB\bar{C}\bar{D} + AB\bar{C}D +$
$ABCD + A\bar{B}\bar{C}D + A\bar{B}CD$.

$\bar{A}\bar{B}\bar{C}D + \bar{A}\bar{B}CD + \bar{A}BCD + \bar{A}B\bar{C}D + \bar{A}BC\bar{D} +$
$AB\bar{C}\bar{D} + AB\bar{C}D + ABCD + A\bar{B}\bar{C}D + A\bar{B}CD$

Finally, let us examine the map depicted in Figure 21-15. In this map there are two groups comprised of two 1s in the third row and four 1s in the second and third column. Remember this is because the map can be thought of as being rolled into a cylinder in which the left and right edges are adjacent, and the top and bottom are adjacent.

We can now apply the grouping concept to the previous developed maps. First, we attempt to group the adjacent squares in the map in Figure 21-11. We find that there are *no* adjacent squares. Therefore, each square is a group standing alone.

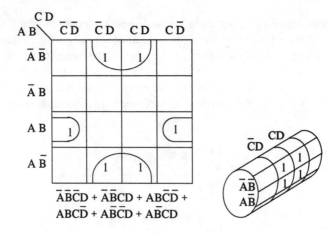

Figure 21–15
A map of groups that are adjacent because of the cylinder effect.

$\overline{A}\overline{B}\overline{C}D + \overline{A}\overline{B}C\overline{D} + A\overline{B}\overline{C}\overline{D} +$
$A\overline{B}C\overline{D} + A\overline{B}\overline{C}\overline{D} + A\overline{B}CD$

Next, let us group the squares of the map developed in Figure 21-12. We observe in Figure 21-16 that there are two groups that are adjacent. Three 1s are in the horizontal group and two 1s are in the vertical group.

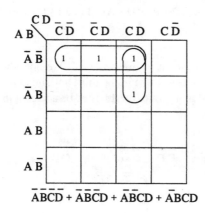

Figure 21–16
Grouping the squares of the map in Figure 21-12.

$\overline{A}\overline{B}\overline{C}\overline{D} + \overline{A}\overline{B}\overline{C}D + \overline{A}\overline{B}CD + \overline{A}BCD$

Having learned to map Boolean equations we can now utilize the maps as a vehicle to reduce equations. *The terms that will appear in the reduction are those that do not change within the group.* Examining the horizontal group in line one of Figure 21-16, we observe that $\overline{A}\overline{B}$ does not change. Next, examining the group in line three we observe that CD does not change. We group the terms that do not change for the reduced equation $\overline{A}\overline{B} + CD$.

Summary of Rules

1. Change the Boolean equation in the form of AND-SUM (AND-OR) by the use of Boolean laws $(ABC + ABC)$.
2. Lay out the KARNAUGH map table in the proper format.
3. Place a 1 in the square representing each AND term of the equation.
4. Group the 1s in the map.
5. List variables of each group that do not change in any square within a group.

Let us apply the rules to another example.

Example 5 Map and reduce the Boolean equation

$$\bar{A}\bar{B}CD + \bar{A}\bar{B}C\bar{D} + \bar{A}B\bar{C}\bar{D} + \bar{A}B\bar{C}D + \bar{A}BCD + \bar{A}BC\bar{D} + ABCD + ABC\bar{D} + A\bar{B}CD + A\bar{B}C\bar{D}$$

to the lowest terms $=\bar{A}B\bar{C} + C$.

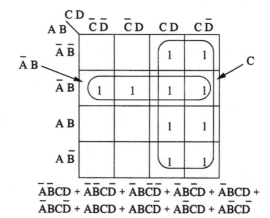

Figure 21–17
Reducing the equation in
Example 5 to the lowest
terms.

$$\bar{A}\bar{B}CD + \bar{A}\bar{B}C\bar{D} + \bar{A}B\bar{C}D + \bar{A}B\bar{C}\bar{D} + \bar{A}BCD +$$
$$\bar{A}BC\bar{D} + ABCD + ABC\bar{D} + A\bar{B}CD + A\bar{B}C\bar{D}$$

After mapping Example 5 we observe that there are two groups of terms: the horizontal group comprised of four 1s and the vertical group comprised of eight 1s. In the horizontal group $\bar{A}B$ does not change and is selected. In the vertical group C does not change and is selected. The ten AND gates in the original equation reduce to $\bar{A}B + C$.

Problems 21–2 Apply KARNAUGH mapping to the following equations to obtain the minimum number of terms.

1. $\bar{A}BCD + A\bar{B}\bar{C}D$
2. $\bar{A}BCD + \bar{A}B\bar{C}D + A\bar{B}\bar{C}\bar{D}$
3. $\bar{A}BCD + \bar{A}B\bar{C}D + \bar{A}\bar{B}\bar{C}D$
4. $\bar{A}B\bar{C}D + \bar{A}BC\bar{D} + \bar{A}\bar{B}CD + \bar{A}\bar{B}\bar{C}\bar{D}$
5. $\bar{A}B\bar{C}D + \bar{A}\bar{B}CD + \bar{A}\bar{B}\bar{C}D + ABCD$
6. $A\bar{B}CD + A\bar{B}\bar{C}D + A\bar{B}\bar{C}\bar{D} + ABCD$
7. $ABC\bar{D} + \bar{A}BC\bar{D} + A\bar{B}C\bar{D} + A\bar{B}\bar{C}\bar{D} + \bar{A}\bar{B}\bar{C}\bar{D}$
8. $\bar{A}BCD + \bar{A}BC\bar{D} + \bar{A}B\bar{C}D + \bar{A}\bar{B}CD + \bar{A}\bar{B}\bar{C}D$

Summary

1. ORing in Boolean algebra applies to the rules of addition.
2. ANDing in Boolean algebra applies to the rules of multiplication.
3. A bar over a variable indicates the inversion, or complement, of the variable.
4. Postulates are obvious relationships.
5. The laws of relationships between terms in Boolean algebra are the *commutative*, the *associative*, and the *distributive laws*.

6. Demorgan's theorems change AND functions to OR functions. The first theorem states that *the complement of the product can be changed to the sum of the complements of each term.*

$$\bar{A}\bar{B} = \bar{A} + \bar{B}$$

7. Demorgan's second theorem states that *the sum of complements can be changed to the complement of the product.*

$$\bar{A} + \bar{B} = \bar{A}\bar{B}$$

8. Boolean equations can be simplified by using Boolean algebra and Demorgan's theorems or KARNAUGH maps.

Answers for Basic Electronics Math

Exercises 1-1

1. 235
3. 724
5. 988
7. 988
9. 346

Exercises 1-2

1. $\frac{7}{6}$ or $1\frac{1}{6}$
3. $7\frac{5}{7}$ or $\frac{54}{7}$
5. 1
7. $\frac{172}{15}$ or $11\frac{7}{15}$
9. $\frac{171}{4}$ or $42\frac{3}{4}$
11. $\frac{1}{4}$

Exercises 1-3

1. 32
3. 54
5. 350
7. undefined
9. 288
11. 15.7

Chapter 2

Exercises 2-1

1. 27
3. 2197
5. 125
7. 441
9. 2401

Exercises 2-2

1. 3
3. 401.3
5. 59049
7. 32
9. 27

Problems 2-1

1. 116.6 W
3. 14.5 A
5. 9 V at 50 mA
7. 12 V, 600 W, 50 A
9. 12 V, 20 A, 0.6 Ω
11. H.L. = 20 A, 0.6 Ω
    D.L. = 0.125 A, 96 Ω
    Dist. = m4.33 A, 2.77 Ω
    R = 5.17 A, 2.32 Ω
    TL = 2.08 A, 5.77 Ω
    H = 7.1 A, 1.69 Ω

Chapter 3

Exercises 3-1

1. 2.7×10^3
3. 1.2×10^{-2}
5. 8.7×10^4
7. 1.2×10^{-5}
9. 1×10^{-8}

Exercises 3-2

1. 10^{-2}
3. 10^2

5. 10^{-5}
7. 1×10^8
9. 1.58×10^7
11. 8.88×10^7

Problems 3-1

1. 7.11 kW
3. 21.7 kW
5. 3 W
7. 0.144 Ω

Chapter 4

Exercises 4-1

1. 1 MΩ
3. 2.56 kΩ
5. 15 GHz
7. 0.22 mH

Exercises 4-2

1. 15.35 in.
3. 91.44 cm.
5. 0.635 cm.
7. 2187 yds.
9. 391 dm.
11. 12.5 cm.
13. 0.0254 mil in.

Problems 4-1

1. 76.2 in.
3. 182.9 yds., 548.7 ft., 2194.7 in.
5. 6780 ft., 2066.5 m.
7. 0.0000406 cm.
9. 0.256 kHz
11. 3 meters, 118.11 inches, 9.84 feet
13. 1.24 GHz
15. 914.4 cm³
17. 0.769 to 1
19. 91.4 meters

Chapter 5

Exercises 5-1

1. 13
3. 8
5. 22
7. 16
9. 23
11. 121

Exercises 5-2

1. 101110011111
3. 1111010
5. 110101111
7. 10101
9. 11100101000

Exercises 5-3

1. 47
3. 511
5. 371
7. 312
9. 110111
11. 111111111
13. 10110101010

Exercises 5-4

1. 10001
3. 11010
5. 11000

Exercises 5-5

1. 1
3. 10
5. 10

Exercises 5-6

1. 101
3. 111
5. −11010
7. 1111
9. 110101

Exercises 5-7

1. 22
3. 350
5. 1516
7. 2027
9. 12
11. 2C
13. D6
15. 3B
17. 256D

Chapter 6

Problems 6-2

1. 176
3. 13.3 yds/sec down
5. 72°
7. 10 lbs left
9. 465.7 mi/hr
11. 373 watt/sec
13. −$116.35 overdrawn

Chapter 7

Exercises 7-1

1. -6
3. $-xy$
5. -60
7. $-5x - 5y$
9. $3x - 3y$
11. $-abcy$

Exercises 7-2

1. -27
3. x^2
5. I^4
7. 8
9. $16a^2c^4$
11. $4b^6$

Exercises 7-3

1. $16x^2y$
3. $2ab^2 - 2ab^3$

5. $16k^2h^2m$
7. $21a^3x^5$
9. $8I_1R_1 - 8I_1R_2$
11. $2xy^2 - 6x^2y^2 - 4x^3y^2$
13. $5ax^3 - 3x^4y$
15. $a^2 - b^2$
17. $h^2 - k^2$
19. $10x^2 + 29xy + 21y^2$
21. $x^5 + 4x^4 + 10x^3 + 13x^2 + x - 2$
23. $2x^4 + 7x^3 + 9x^25x + 1$
25. $2I_1R_1$
27. $a^{2n} + a^{2n+2}b^2 + a^{3n-3} + a^5b^3$

Chapter 8

Exercises 8-1

1. a
3. 3
5. x^{-4}
7. I^{-5}
9. 2
11. 7^2
13. r^2
15. E^{-5}
17. b^5
19. $1/A^5$
21. s^6
23. R
25. $3^{3/2}$
27. $8^{1/2}$
29. 2^3

Exercises 8-2

1. a
3. $0.5\ \pi R$
5. $0.5I$
7. x^{a-b^1}
9. $1/3$
11. $1/2\ \pi w$
13. 1
15. $a^{-1}b$
17. $2/(x^{3/2}y^{5/2})$
19. $2N_1^{-1}$

Exercises 8-3

1. ax^2
3. $.5x^2 + x + 1.5$
5. $2 + r$
7. $.5ab - c^2$
9. $h^2 r^{-1} + \frac{1}{6} h$
11. $r_1/r_2 = 1 + r_2/r_1$
13. $I_1 R_1 R_2 / I_2 - I_2 R_2^2 / I_1$

Exercises 8-4

1. $3a + 2(R)5$
3. $x + 8$
5. $x + y$
7. $x - y$
9. $2^2 - 3ER + R^2$
11. $x - 2y$
13. $R^2 + 0.3R + 2.56$
15. $-6.5I + 13.75$

Chapter 9

Exercises 9-1

1. 5
3. -1
5. -4
7. 7
9. $9\frac{1}{2}$
11. $\frac{1}{2}$
13. 4
15. 25
17. 4
19. $R = 3, -1$
21. $a = 1$
25. $p = -\frac{1}{2}$
27. $n = -12$

Exercises 9-2

1. 300 sq. ft.
3. $0.125\ \Omega$
5. 100 ft. sq.
7. 500 kHz
9. 180 cu. cm.
11. 0.02 coulombs
13. $-4V$
15. 0.5 power factor

17. E/I
19. $\pm\sqrt{z^2 + x^2}$
21. Q/E
23. $N_2 \sqrt{z_p / Z_S}$
25. $\dfrac{R}{pf}$
27. $\dfrac{R_t}{1 - \sigma t}$
29. $\dfrac{pL}{R}$
31. $\dfrac{HL}{N}$
33. $G_t - G_1 - G_2$
35. $\dfrac{X_L}{Q}$

Chapter 10

Exercises 10-1

1. $x(ax + 1)$
3. $2IR(R + 2I)$
5. $2xy^2(2xy + 1)$
7. $2\dfrac{a}{b}\left(a - \dfrac{2}{b}\right)$
9. $\pi^2 r^2$
11. $a^2 b^2 c^4$
13. $9x^4 y^2$
15. $\dfrac{25I^2}{R^4}$
17. $x^4 y^4 z^6$
19. $\dfrac{25}{I^4}$
21. $\dfrac{4\pi^2}{9T^2}$
23. $-\dfrac{4\pi^4}{25E^2}$
25. $8\pi^3$
27. $-8\pi^3 r^3$
29. $\dfrac{a^3}{p^6}$
31. $-8x^9 y^3$
33. $\dfrac{8h^6}{27}$

35. $-\dfrac{8c}{27}$

37. $\dfrac{i^3 R^6}{T^3}$

39. IR

41. $2\pi h$

43. $a^{1/2} b c^{1/2}$

45. $-\pi/R^2$

47. $\dfrac{2ab}{x^2 y^3}$

49. $\dfrac{6R^3}{5p}$

51. $\dfrac{x^3 y^4}{4z}$

53. -3

55. $^3/_2$

57. $-^1/_2$

59. $2xy^3$

61. $3xy^2$

63. $(6R)^{1/3} 2RI^2$

65. $\dfrac{15}{r}$

67. $\dfrac{9RI^2}{2mn}$

69. $\dfrac{\pi^{1/9} r^{1/6}}{B^{1/6}}$

Exercises 10-2

1. $2a^2 x^2 + 2a^2 y^2$
3. $E^2/R_2 + E^2/R_3$
5. $a^2 - 2ab + b^2$
7. $x^2 - y^2$
9. $a^2 + 2ab + b^2$
11. $E^2 - 10E + 25$
13. $R^2/4 + R + 1$
15. $a^6 y^4 + 2a^3 b^2 y^2 + b^4$
17. $R^2 I_1^2 + 2r^2 I_1 I_2 + R^2 I_2^2$
19. $25R^2 + 10RIE + I^2 E^2$
21. $4R^2 - 9I^2$
23. $x^{10} - 14x^5 + 49$
25. $4B^2 + 20B + 21$
27. $0.01R^2 - 0.25RZ + Z^2$
29. $2x$
31. 9
33. θ^2
35. $12ZR$

37. $(x + 2)^2$
39. $(Z - 2)(Z + 5)$
41. $(Z - 5)(Z + 2)$
43. $(x - 4y)(x - 9y)$
45. $(W - 1/8)^2$
47. $4R(I + E)^2$

Exercises 10-3

1. $a^2 + ab - 2b^2$
3. $2I^2 + 3IR + 2R^2$
5. $I^2 + 2IR + 15R^2$
7. $4x^2 + 24x + 35$
9. $a^2 - ab - 2b^2$
11. $2n^7 - 5n^5 + 22n^2 - 55$
13. $\phi - \theta^2$
15. $2t^2 - Tt - 6T^2$
17. $0.02v^2 + 0.17vw - 0.3w^2$
19. $0.9r^2 + 1.14ri - 3.36i^2$

Exercises 10-4

1. $(I + R)^2$
3. $E^2 \pm 3I$
5. $R \pm 1$
7. $4t^2 \pm 2\delta$
9. $\left(\dfrac{1}{3}a - \dfrac{1}{5}b\right)$
11. $(5I - 9)$
13. $(I^3 - 4IR^2)(I^3 + 4IR^2)$
15. $[R + (E - I)][R - (E - I)]$
17. $(a/3 + b/4)$
19. $(E/R + 1/9)$

Exercises 10-5

1. $(x + 1)(x - 3)$
3. $(5R - 9E)(5R - 9E)$
5. $(I - Z)(R - E)$
7. $(1 + 2x)(1 + 2x)$
9. $(a - 4b)(a - 7b)$
11. $(2x - 3)(2x - 3)$
13. $(2a^n - 5)(2a^n - 5)$
15. $(2w + 9v)(2w + 5v)$
17. $(7 + E)(3 + E)$
19. $(2I - 3E)(2I - 3E)$
21. $(w - 11v)(w - 7v)$
23. $(a - b)(x + y)$

25. $(R + E - 5)(R + E + 6)$
27. $(a - b)(a - b - 2)$
29. $(2Z + 3)(Z + 1)$
31. $(3R - 5)(a - 2b)$
33. $(5N - 7)(N + 2)$
35. $(3)(a)(a)(a)(1 - 9b)(1 - 4b)$
37. $(x - y - i)(x + y + i)$
39. $(3)(3)(E - 2 - I)(E - 2 + 1)$
41. $(I + R - 3)(I - R + 3)$
43. $(2)(2)(a)(7x - 1)(x + 2)$
45. $(E - R - 1)(E + R)$
47. $(e + i - 1)(a - b)$
49. $(a + e)(a)(e - 1)$

Chapter 11

Exercises 11-1

1. 6
3. 8
5. 6
7. 6
9. 3
11. 5

Exercises 11-2

1. 8
3. a^3b
5. $a + b$
7. I
9. $e + 2$
11. $I - r$
13. No common factor

Exercises 11-3

1. 64
3. $24a^2b^2c^2$
5. aI^3R^2E
7. $(a + b)^2(a - b)$
9. $4(I - 1)(I + 2)(I - 3)$
11. $(P + 2)(P - 2)(P + 3)(P - 3)$
13. $12ev^2(v + e)^2(v - e)^2$

Exercises 11-4

1. $\dfrac{26y^3}{37}$

3. $\dfrac{R_2 + R_3}{E + I}$

5. $\dfrac{x + y}{x - y}$

7. $\dfrac{e + i}{e - i}$

9. $\dfrac{1 + 2I - 3I^3}{3RI - 4I + 1}$

11. $\dfrac{x + y}{2}$

13. $\dfrac{R + 4}{R - 4}$

15. $a - b$

17. $\dfrac{b + 2a}{b - a}$

Exercises 11-5

1. $\dfrac{a^3 + 1}{a^2}$

3. $\dfrac{4a - 3}{1 - a}$

5. $\dfrac{\pi h - cb - 3c}{b + 3}$

7. $\dfrac{-2B - 3}{B + 1}$

9. $\dfrac{R_1 + 2 + R_2W + W}{W}$

11. $\dfrac{E_1 + E_1R_1}{R_1}$

13. $\dfrac{3a^2 + a^2b + 2a - 2 - a^2 - b}{a^2 - 1}$

15. $\dfrac{3r^2 + r - h - hr}{1 + r}$

17. $\dfrac{3\theta\lambda - \phi\theta\lambda - \phi w - 2w}{(\theta + 2)\lambda}$

19. $\dfrac{\Omega n + \Omega\lambda + \Omega\rho + \lambda\rho + \rho\alpha n + \rho\alpha\lambda}{\alpha(n + \lambda)}$

21. $\dfrac{1}{a} + \dfrac{2}{a^4}$

23. $\dfrac{a}{1 - a} + \dfrac{2}{1 - a}$

25. $R_1 + \dfrac{R_1}{E_1}$

27. $\dfrac{x}{x-1} + \dfrac{1}{x-1}$

29. $\dfrac{\alpha^2}{1-\alpha} + \dfrac{\alpha\lambda}{1-\alpha} + \dfrac{\lambda^2}{1-\alpha}$

31. $\dfrac{(\theta-1)(\theta+1)}{(\theta-1)(\theta+1)}$

33. $\dfrac{2\tau}{2\tau-a} - \dfrac{\tau^3}{2\tau-a} - \dfrac{\tau^5}{2\tau-a}$

Exercises 11-6

1. $\dfrac{a^2x+1}{a^2}$

3. $\dfrac{1+xy^3}{xy}$

5. $\dfrac{2ai^2 - 2aiv + i^2 - 2iv + v^2 + 2av^2}{a^2+a}$

7. $\dfrac{R_2V_1^2 - 2V_1^2 - R_1V_2^2 - 3V_2^2}{(R_1+3)(R_2-2)}$

9. $\dfrac{-16I-9}{72I}$

11. 4

13. $\dfrac{a^2 - 3ax + x^2}{a(a+x)(a-x)}$

15. $\dfrac{-5}{6(Z+1)(Z-1)}$

17. $\dfrac{a^2 + 2a + 1}{a}$

19. $\dfrac{10x + 3y}{(3x-y)(3x-y)}$

21. $\dfrac{-2ax^2 - 7ax - a}{x(x-1)(x+1)}$

23. $\dfrac{-13a^2 + 8a - 8}{24a^2}$

Exercises 11-7

1. $\dfrac{x}{a^2}$

3. $6IR2$

5. $X^2 - 1$

7. $\dfrac{i^2 - v^2 + v^2i + vi^2}{i^2v^2}$

9. $\dfrac{a^4b + ab - a^3b - b}{a^4 - a^3 + a^4b - ab}$

11. $\dfrac{R^2 - 14R - 15}{R^2 - 25}$

13. $\dfrac{2a}{x - 2y}$

15. $4x^3y^3z$

Problems 11-1

1. 11 V
3. 5 Ω
5. 187.5 Ω
7. a. 38792
   b. 686
   c. 2134
9. 179.2 sq. cm., 355 sq. cm., 2550 sq. cm.

Chapter 12

Exercises 12-1

1. $E = 4$
3. $R = {}^{24}/_5$
5. $Z = 5$
7. $\beta = -6$
9. $^1/_6 = c$
11. $n = 3m - 3$
13. $E = {}^3/_{26}$
15. $\phi = {}^{14}/_{15}$

17. $R_T = \dfrac{1}{\frac{1}{R_1} + \frac{1}{R_2} + \frac{1}{R_3}}$

19. $L = \dfrac{1}{4\pi^2 fo^2 C}$

21. $N = \sqrt{\dfrac{\phi}{1.261}}$

23. $F = {}^9/_5\, C + 32$

25. $h = \dfrac{A - 2\pi r^2}{2\pi r}$

27. $m_1 = \dfrac{Fr^2}{m_2}$

29. $r = \dfrac{V}{\pi h^2} + \dfrac{h}{3}$

31. $g = \dfrac{2(S - Vot - so)}{t^2}$

33. $L = \dfrac{R(Z-1)}{W}$

35. $f = \dfrac{x}{n+1}$

37. $q = \dfrac{\mu KT}{D}$

39. $R_L = A_v r_{eb}$

43. $f = \dfrac{300 \times 10^{-6}}{\lambda}$

45. $R_p = \dfrac{z_1 e_1}{\mu eg - ei}$

47. $\beta = \dfrac{\phi\theta\lambda}{\theta\lambda + \phi\lambda + \phi\theta}$

49. $R_B = \dfrac{R_E - SR_E}{S(1 - \infty) - 1}$

Problems 12-1

1. $396 \ \Omega$
3. missing part of figure 12-7
5. $1822 \ \Omega$

Chapter 13

Problems 13-1

1. $V_L = 6.38 \text{ V}, V_{th} = 9.57 \text{ V},$
   $R_{th} = 34.3 \ \Omega$
3. $R_{th} = 5.2 \text{ k}\Omega, G_n = 0.833 \text{ mS},$
   $I_N = 11.5 \text{ mA}, V_L = 39.5 \text{ V}$

Chapter 14

Exercises 14-1

1.

3.

5.
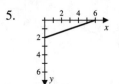

Exercises 14-2

1. $V_L = 2.84$, electron flow upward
3. $R_{th} = 63.8 \ \Omega, V_{th} = 20.64 \text{ V},$
   $V_L = 6.14 \text{ V}$
5. $V_1 = V_2 = 5 \text{ V}$

Problems 14-1

1. 10, 30, and 10
3. .4 minutes, .4 minutes, and
   .10 minutes
5. Receivers = $150 each,
   Albums = %5 each and
   20 Phonographs.
7. $361 \ \Omega$
9. 24 resistors, 40 capacitors,
   and 20 fuses.

Chapter 15

Exercises 15-1

1. -1 at $-90°$
3. $+1$ at $+90°$
5. -1 at $-90°$
7. -12 at $-90°$
9. 4 at $0°$
11. -3 at $-90°$
13. -4 at $-90°$

Exercises 15-2

1. $7\sqrt{3}j$ or $j7\sqrt{3}$
3. $j16$
5. -2
7. 5
9. $12 + j3$
11. -12
13. -6

15. $3\sqrt{5}$
17. -30
19. -1
21. $j4$
23. $j12\sqrt{3} - j6$
25. -8
27. $j3$
29. 0.5
31. $\dfrac{\sqrt{5}}{\sqrt{2}}$
33. j
35. $-j2.5$
37. -0.6
39. -0.5

Problems 15-1

1. $90°$ rotation or $\sqrt{-1}$
3. yes!
5. $2 - \sqrt{3}$

Chapter 16

Exercises 16-1

1. 0.60206
3. 1.60206
5. -3.9794
7. -3.52287

Exercises 16-2

1. 20 dB
3. 120 dB
5. 55.56 dB
7. 35.6 dB
9. 73.9 db

Problems 16-1

1. 1.59 kHz
3. 1.MHz
5. 15.9 kHz

Problems 16-3

1. $f_1 = 6.36$ Hz, $f_2 = 48$ kHz,
   $f_m = 5.5$ kHz, $BW = 47.4$ kHz,
   $A_{vm} = 0.66$

3. $A_{vm} = 0.8$, $f_1 = 32$ Hz,
   $f_2 = 2$ MHz, $BW = 2$ MHz,
   $f_{mid} = 23.5$ kHz
5. $f_1 = 80$ Hz, $f_2 = 580$ kHz,
   $BW = 580$ kHz, $f_m = 2.15$ kHz

Problems 16-2

1. 14.77 dB
3. 0.792 dB
5. a. $+31.6$ db, b. -31.6 dB
7. 0.5 mW
9. 54 dB

Chapter 17

Problems 17-1

1. $a = 2, b = 2.25, c = 2.38, d = 2.44,$
   $e = 2.49, f = 2.71$
3. -2.3 μsec.
5. 0.105 RC
7. 142 μsec.
9. 7.05 kHz
11. 2.2 TC

Problems 17-2

1. 10 nsec.
3. 200 nsec.
5. 800 watt-sec.
7. 4,900 joules
9. 1

Chapter 18

Exercises 18-1

1. 0.48 rad.
3. 1.31 rad.
5. 5.5 rad.
7. 3.03 rad.
9. 360°
11. 137.5°
13. 28.65°

Exercises 18-2

1. sin = 0.866, cos = 0.5, tan = 1.732
3. sin = 0.707, cos = 0.707, tan = 1
5. sin = 0.5, cos = 0.866,
   tan = −0.577
7. sin = −0.707, cos = −0.707,
   tan = −1
9. sin = −0.866, cos = −0.5

Chapter 19

Exercises 19-1

7. 10 msec, 628t, −270°
9. 19.2 Hz, 52 msec., 120.5t, 120°

Exercises 19-2

1. 50 + j86.7
3. 0 + j100
5. 100 + j0
7. 77.9 + j62.7
9. 54.4 + j54.4
11. 49.5 + j65.5

Problems 19-1

1. 603 mph
3. 20.39 knots at 56.3°
5. 151.1 miles above earth
7. 304.2 lb. at 25.3°

Chapter 20

Problems 20-1

1. −118 V
3. 212 V
5. 28.5°

Problems 20-2

1. 68.4 mA
3. 12 MΩ
5. 158.4 μA
7. 339.4 V

Problems 20-3

1. 13.23 W
3. 932 μW
5. 4.53 A
7. 50 μA

Problems 20-4

1. 0.089 μA
3. 5 pF

Problems 20-5

1. 820/75.9°
3. 0.156 A
5. 0.64 mA, θ = −57.8°

Problems 20-6

1. 79.6 mA
3. 802 Ω, 249 μA
5. R = 16.3 Ω, L = 3.47 Ω

Problems 20-7

1. 0.0047 μF
3. 11

Exercises 20-1

1. R = 69.2 kΩ, X_c = 4.61 kΩ,
   θ = −33.7°
3. R = 2.6 kΩ, X_c = 1.5 kΩ, θ = −2°
5. R = 2.6 kΩ, X_c = 1.5 kΩ, θ = −30°
7. R = 1.99 kΩ, X_c = 642 Ω,
   θ = −17.9°

Exercises 20-2

1. R = 69 Ω, X_c = 46 Ω
3. R = 4.33 kΩ, X_L = 7.5 kΩ
5. R = 1.51 kΩ, X_L = 1.41 kΩ
7. 64 Ω
9. 277 Ω, 872 Ω

Exercises 20-3

1. $f_r = 15.9$ kHz, $f_1 = 14.9$ kHz,
   $f_2 = 16.9$ kHz, $BW = 2$ kHz
3. $f_r = 159$ kHz, $BW = 3.2$ kHz,
   $Q = 50$, $F_1 = 157$ kHz,
   $f_2 = 160, 6$ kHz

Problems 20-9

1. $Z = 260/\underline{4.4°}\,\Omega$
3. $Z = 19.6\,\text{k}/\underline{-90°}\,\Omega$
5. $Z = 4.25\,\text{k}/\underline{-45°}\,\Omega$

Problems 20-9

1. $C = 45.5\ \mu\text{F}$
3. $F_r = 12.6$ kHz
5. $R_p = 19.6\ \text{k}\Omega$

Problems 20-10

1. 0
3. A.
5. At the output

Chapter 21

Exercises 21-1

1. A+B
3. ABC
5. AB+C
7. A+BC

9. $A\bar{B}+AC$
11. $\bar{A}B+C$
13. C
15. $A+\bar{B}$
17. $AB+A\bar{C}$
19. A+CD

Exercises 21-2

1. $\bar{A}+AC$
3. ABC+ABD
5. AB(C+D)
7. AC
9. ABC
11. $\bar{A}B\bar{C}$

Index

Decimal numbers (*cont.*)
    conversion
        from binary, 27
        to binary, 27–28
        table, 30
    placement identification, 1 (table)
Decoding, 27
Degrees
    of angles, 144
    of monomials and polynomials, 79
Demorgan's theorems, 183–184
Denominators, 2
    of algebraic fractions, 82–83
    common, 2–3
    conjugates of, 122
    lowest common. *See* LCD.
    moving factors to, 60
    rationalizing, 122–123
Determinants, 115–118
Digital electronics. *See* Electronics.
Directed magnitude, 122
Direction (vectors), 150
Discharge of capacitor, 138
Dissimilar terms, 38
Distributive law, 181
Division, 60
    algebraic fractions, 88–90
    logarithms, 125, 127
    monomials, 55–57
    polynomials by monomials, 57–58
    polynomials by polynomials, 58–60
    powers of ten, 13–15
Drivers, 178
DVM, 157

Effective current, 156. *See also* Current.
Effective voltage, 156. *See also* Voltage.
Eight's complement, 34
Electrical energy, 37
Electron volts, 18
Electronics
    binary numbering system use by, 27
    measurements in, 22
    prefixes, 19 (table)
    units for, 18–19
    use of logarithms in, 137
Encoding, 27–28
End-around carry, 32
Energy, 37
English statements, translating into mathematical
        formulas, 65–66
English system of measurement, 20 (table)
    compared to metric system, 22–23
Epsilon, 137
Equal factors, 8, 9
Equations, 37–38
Equations. *See also* Expressions; Formulas.
    algebraic, 37–38
    axioms for solving, 61–63
    Boolean, 183, 185–189
    fractional, 93–97

Equations (*cont.*)
    indeterminate, 109–115
    linear, 112 (fig.), 115
    product-sum, 185–186
    rules for solving, 63–65
    simultaneous, 108–115
Equilateral triangles, 146
Equivalent circuits, 102–105, 133 (fig.). *See also*
        Circuits.
Equivalent fractions, 82
Equivalent series circuits. *See* ESC.
ESC (equivalent series circuits), 166, 172
Evolution, 39–40, 125, 127
Exactness, 65
Exponential formulas
    charging a capacitor, 138
    inductance and, 140–142
Exponential functions, example, 138 (fig.)
Exponents, 6, 11, 48, 60
    laws of, 15
    multiplying monomials with, 49
    multiplying signed numbers with, 48–49
    negative, quantities with, 60
    in quotients, calculating, 54
    signs of, 11–12
    subtracting, 53–55
    use in division by powers of ten, 13
    zero as, 60
Expressions. *See also* Equations; Formulas.
    algebraic, 37–38
    mathematical, 4
    mixed, 84–85
Extraction of
    cube roots, 69–70
    square roots, 8, 69

Factoring, 68–70, 71–72
Factors, 37, 68. *See also* Prime factors.
    common, 79, 83
    equal, 8, 9
    highest common. *See* HCF.
    monomial, of polynomials, 71–72
    of trinomials, 76–78
Fahrenheit scale, 42
Faraday's law of electrolysis, 67
Filter circuits, band-pass, 131–135
Filters, 130–131
Flux density, 66
Formulas, 36, 37–38. *See also* Equations; Expressions.
    exponential, 140–142
    indeterminate, 109
    translating English statements into, 65–66
Fractional equations, 93–97
Fractions
    adding, 2–3
    algebraic, 82–90
    clear of, 95
    common, 2
    complex, 88
    converting, 84–85
    decimal, 1

Fractions (*cont.*)
   equivalent, 82
   improper, 2
   roots of, 8
   subtracting, 3
Frequency
   cutoff, 129, 134, 170
   discrimination, 129
   in alternating current, 155–156
   of a radius vector, 151
   parallel resonant, 174, 175
   response, 129–130
Functions. *See* Circuits, functions; Exponential
      functions; Periodic functions; Trigonometric
      functions.

Gain of a transistor, 67
Geometrical representation
   of complex numbers, 121 (fig.)
   of imaginary numbers, 119–120 (fig.)
   of real numbers, 120 (fig.)
Grams, 20
Graphical solutions
   simultaneous equations, 109–111
   vector calculations, 153
Graphs, 108–111
Ground, 42, 47
Grouping
   KARNAUGH maps, 186–189
   signs of, 4–5

HCF (highest common factors), 79–81
Hertz, 19, 151, 155
Hexadecimal numbering system, 26, 29
Hexadecimal numbers
   adding, 31
   conversion table, 30
High-frequency, 133 (fig.), 135
High-pass filter, 131, 134
Highest common factors. *See* HCF.
Horizontal coordinate. *See* Abscissa.
Hypotenuse, 145
Hz. *See* Hertz.

Imaginary numbers
   geometrical representation of, 119–120
   pure, 120–121
Impedance, 66, 160
   of RC circuits, 166, 167
   of series circuits, 166–167
   of series-parallel circuits, 169–170, 171–173
   phase angle, 166
   reciprocal of, 164
   triangular form, 153 (fig.)
Improper fractions, 2
In phase, 157
Indeterminate equations, 109
Indexes, of roots, 9
Inductance, 164
   conducted in parallel, 167, 168 (fig.)
   connected in series, 162–163

Inductance (*cont.*)
   exponential functions and 140–142
Inductive circuits, 140–142. *See also* Circuits.
Inductive reactance, 66, 162, 167. *See also* Reactance.
Initial side, 143
Input, 177–178, 180–181
Instantaneous voltage, 155–156. *See also* Voltage.
Internal battery resistance, 67
International System of Weights and Measures, 20. *See*
      also SI metric system of measurement.
Inverse operations, 7
Inverters, 178 (fig.)
Involution, 125
Irrational numbers, 119
Isoceles right triangle, 147 (fig.)

j operator notation, 164–165, 167–169, 173
Joules, 37

KARNAUGH maps, 185–189 (figs.)
Kelvin scale, 42
Kirchhoff's current law, 39
Kirchhoff's law, 99
Kirchhoff's voltage law, 49, 159

LCD (lowest common denominators), 3, 85–87, 93. *See*
      also Denominators.
LCM (lowest common multiples), 81–82
LCR circuits, 169–171. *See also* Circuits.
Least significant digit. *See* LSD.
Length, common metric measurements of, 21 (table)
Like terms, 38
Linear equations
   determinants used to solve, 115
   solving circuit currents by, 112 (fig.)
Literal coefficients, 38
Literal parts, of terms, 38
Load resistance, 129
Logarithmic center, 134
Logarithms. *See also* Antilogarithms.
   applications of, 125, 127–128
   common, 125–126, 127–128
   division by, 125, 127
   evolution, 125, 127
   involution, 125
   multiplication by, 125, 126–127
   natural, 125, 137–140
Logic diagrams, 182 (fig.), 183 (fig.)
Logic expressions, 178–179
Logical algebra, 177
Low-frequency, 133 (fig.), 135
Low-pass filter, 131, 134
Lowest common denominators. *See* LCD.
Lowest common multiples. *See* LCM.
LR circuits, 141. *See also* Circuits.
LSD (least significant digit), 1–2

Magnetic field intensity, 67
Magnitude (vectors), 150, 152, 154
Mathematical expressions, 4
Mathematics, definition, 65